THE TECHNOLOGY CENTURY

100 Years of ESD — The Engineering Society 1895-1995

Mike Davis, editor

Cover, frontispiece and section front illustrations:
Detroit Industry, South Wall, fresco by Diego Rivera
(Detroit Institute of Arts; gift of Edsel B. Ford)

CENTENNIAL MISSION STATEMENT OF *ESD — THE ENGINEERING SOCIETY:*

In 1895 the Association of Graduate Engineers was established in Ann Arbor, Mich. to support and promote the growth of the engineering and technology based industries by providing the engineering community with opportunities to share knowledge and ideas. The Society's Centennial Celebration will commemorate and enhance awareness of 100 years of achievements and contributions by its members and the engineering profession, and continue support of its founding purposes with programs and activities to encourage and stimulate economic growth and scientific discoveries in the 21st century.

Contents

Contents

Congratulations

and

Happy Anniversary

THE ENGINEERING SOCIETY 1895-1995
A CENTURY OF ACHIEVEMENTS
A NEW MILLENNIUM OF DISCOVERY

We applaud your achievements, efforts and goals

as the premier society for all engineers.

We are honored to have served your members for

over 20 Years!

Best Wishes from all of us at

Michigan Benefit Plans
P.O. Box 215080
Auburn Hills, MI 48321

1-800-682-6881

Providing Group Insurance Discounts for
Automobile, Homeowners, Workers Compensation, Blue Cross/Blue
Shield, Disability Income Insurance.

Presidents' messages

From John G. Petty

"Where there is no vision, the people perish" (Proverbs 29:18)

As the 21st century dawns and ESD — The Engineering Society embarks on its second 100 year journey, the future will be as bright as we can imagine if we each take personal responsibility today to make it so.

Let's not accept the current premise that our offspring will have a lower standard of living and be less secure in their old age than their predecessors. These visionless predictions will only come true if we continue trying to solve tomorrow's challenges with out-dated or stopgap measures of yesterday.

We can all be very gratified to be a part of the world of engineering, in which the end results are "tangible." It's these tangible, productive results that have perpetuated "the good life."

On the other hand, we have seen in recent years a variety of somewhat "intangible" business activities and government programs erode the foundations of what we view as that "good life."

However, I must also note that there are personal factors that might be viewed as intangible that have made and will continue to make a significant contribution to our growth, such as our spiritual/moral values, imagination and faith.

We live in a complex world and, therefore, the solutions to the challenges facing us are equally complex. Complex projects have been very effectively engineered using the "team" (not committee) approach. Therefore, it would appear that at least part of the answer might be in the application of teams to develop social solutions.

The Engineering Society, consisting of a variety of disciplines and industries, is ideally suited to act as a catalyst in marshaling action-oriented, interdependent teams comprised of very capable members from ESD and our sister and brother affiliate societies to help in solving a myriad of challenges facing us.

These teams will need leaders with vision and participants who have a desire to help implement tangible solutions today for the benefit of tomorrow.

It is my vision that all ESD members will make a valuable contribution to the future of our children and grandchildren; but in order to do so, each one of us must become actively involved today.

John G. Petty
ESD President-elect 1995-96

From John Rakolta Jr.

What an important year for the ESD — The Engineering Society: its 100th anniversary. And what an honor it has been to serve as president of this great organization at such a moment in its history. Centennials always give rise to reflection not only on any great institution's own history but on its place in the events of its time. In pausing to reflect on ESD's role in this region's and even broader history, I am equally awed by past accomplishments and by those poised for realization now and in the future.

This centennial book, this collaborative document — describes and depicts the sweeping achievements attributable to engineering and technology advances in the last 100 years. To see, in words and pictures, the unfolding of the conditions of modern life — many of which we hardly notice anymore — instills both pride in the progress we have made and humility at the powers now at our command.

This dual nature of respect for the past and present underlies the attitude with which we must look to the future. We need to embrace the future secure and inspired by what has gone before. However, we must take care not to succumb either to complacency over our current capabilities nor to discouragement that past prowess can't be matched. We should not see ourselves in a heat to outpace our forerunners, but as holders of the baton of progress that has passed to us, to tirelessly bear for the next lap of the common pursuit.

In that effort, we must breathe deeply the spirit of inspiration, for that is the source — the additional ingredient almost from outside ourselves — that enables us to see outside the ordinary, the limited, the "box" of what is, to what could be. With inspiration we and our descendants may do great things that ESD members will look back on with respect and awe 100 years from now, as we do today. But, most importantly, they will, as we do now looking at our history, become inspired to create anew.

John Rakolta Jr.
ESD President 1994-95

WE TAKE THE PERFORMANCE OF OUR PEOPLE VERY SERIOUSLY.

Although you expect exceptional performance from Chrysler Corporation vehicles, you should also expect it from our people. That's why we've developed a program called Customer One. A renewed commitment to customer satisfaction at each level of the Chrysler Corporation. A commitment that involves everyone from management to the factory to the dealership. A commitment we take very seriously, to help make our services as good as our products. Customer One. It's a new way of looking at our performance. Put it to the test.

Divisions of the Chrysler Corporation.

CustomerOne ☆

Foreword

When I was asked to submit a proposal for ESD — The Engineering Society's centennial publication, the first thing I did was peruse the annual directory issue of ESD Technology to see who the members were. Because the first objective was to serve ESD's constituency.

The second task was to tell about Detroit's technologies and industries during the world's most momentous 100 years.

The third requirement was to consider how future generations would react when they pick up this book 10, 50 or a 100 years from now and try to understand what was going on here in 1995.

What I outlined would first relate the history of the organization whose hundredth year was being celebrated. Of course, any centennial of a Detroit organization had to recognize that this was, and is, Motor City, the home of the auto industry as the mass production enterprises which put first the nation and then the world on motorized wheels. An industry that was celebrating its own centennial a year hence. Then, what did the auto industry cause to happen around it, as it grew? There was the infrastructure — the people, the builders, the suppliers of services to support those people and the factories.

But Detroit has been far more than just the auto industry. There were many industries that flourished here, gained worldwide influence, some still present, others long gone. Their stories had to be told, both to remind the readers of 1995 and to inform those in the future. Conventional histories tend to be either anecdotal memories or recordings of political events. Stories of enterprises seldom seem to be included. The ESD anniversary provided an ideal platform for these annals of The Technology Century.

Finally, a centennial publication should offer a view of the future. Since any student of technology knows that inventions can't be predicted, the challenge was to avoid looking ridiculous even a few months later, when predictions could fall by the wayside as completely unforeseen developments leap ahead. It seemed appropriate, given ESD's founding by University of Michigan College of Engineering graduates, to assemble a trio of faculty members to ponder the future of today's technological community.

After my proposal was approved by Chris Dyrda's publications committee, I was handed a small book from the ESD library. It was a 50th anniversary book published by the Cleveland Engineering Society (now defunct) in 1930. Perhaps not surprisingly, its format followed very much that laid out for this volume.

Quite another aspect of creating our anniversary book was assembling competent writers and historians with a very short lead time to "Job One." Fortunately I could draw upon the resources of journalists as well as amateur and professional historians with whom I had become acquainted over my years as both journalist and historian. In many instances, I was able to identify contributors with unique knowledge of their subjects.

The challenge was to mold professional journalists into competent historians, because history requires time devoted to research, a luxury rarely afforded writers facing hourly or daily deadlines. By the same token, it was necessary sometimes to trim the exhaustive narratives of subjects dear to the hearts of historians.

The contributions of those journalists engaged as public relations professionals generally required the least nurturing, once they could be persuaded that this was a project worthy of their employer's time. Some of the PR efforts were staff-produced and, accordingly, are anonymous.

Because several writers provided more than one article, we decided to group their biographical sketches together in a special section, rather than appending them to each contribution.

When ESD's century began in 1895, it was a world in which electricity and telephones were just beginning to be employed, a world without cars and planes and radios and movies and television, a world where the mechanical typewriter was a relatively new invention.

It was a world in which Detroit was just another medium-sized, regional city — like most northern cities, a place of European immigrants and foreign-language churches, neighborhoods and newspapers. A simpler world in many respects, but one filled with terrible insecurities balanced by unlimited opportunities.

The subsequent 100 years for Detroit truly was The Technology Century. We wonder what our next 100 years will hold, and how it will be characterized.

—Mike Davis, editor-in-chief

ONLY YOUR MOTHER IS MORE OBSESSED WITH YOUR SAFETY.

Ford Safety Engineers: Karin H. Przybylo, Steve Pingston, Mike Foster.

Where would we be without our mothers? They take care of us and protect us. So, we're proud to say, when it comes to safeguarding drivers, at FORD MOTOR COMPANY our maternal instinct becomes very apparent. You can feel it in our TRACTION CONTROL system. And in our ANTI-LOCK BRAKES. It's why DUAL-AIR BAGS* are standard in all our cars. And why ROADSIDE ASSISTANCE is available 24 hours a day. We're also developing a Vision Enhancement System — to help drivers when "mother" nature acts up. All this might be considered obsessive. But at Ford Motor Company, we believe such commitments to safety and security will enhance the quality of all our lives. Besides, it's for your own good.

· FORD · FORD TRUCKS · · LINCOLN · MERCURY ·

QUALITY IS JOB 1 ℠

*Always wear your safety belt.

I. Past: Not Just Prolog

*Detroit area's innovators make it
a microcosm of the industrial age*

General Motors.

"The reasons we believe in Daytime Running Lamps are obvious."

You're driving in broad daylight. Suddenly, a car's lights catch your eyes. They could be our new daytime running lamps (DRLs). These special low-intensity headlights help alert other drivers with light, like your car horn can alert them with sound, according to Jay Minotas, a member of General Motors' safety team. They're simple, practical, and easy to use because they come on automatically. And research shows they reduce collisions, and that can help save lives. GM is introducing daytime running lamps in 1995, and they'll be shining on all its vehicles by 1997. Right now, no other domestic auto company offers them. Jay says he hopes that will change: "The sooner every car and truck has DRLs, the safer we'll all be."

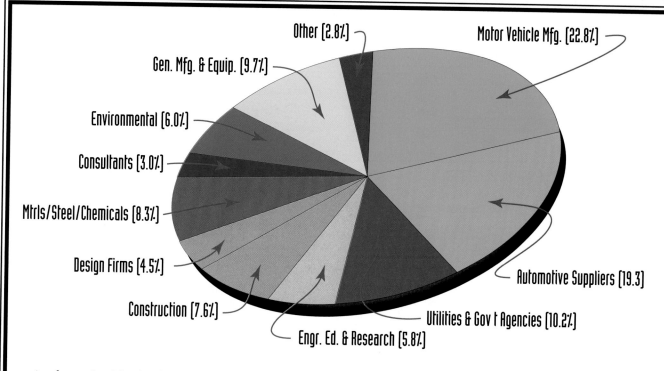

Other [2.8%]　Motor Vehicle Mfg. [22.8%]

Gen. Mfg. & Equip. [9.7%]

Environmental [6.0%]

Consultants [3.0%]

Mtrls/Steel/Chemicals [8.3%]

Design Firms [4.5%]

Construction [7.6%]

Engr. Ed. & Research [5.8%]

Utilities & Gov't Agencies [10.2%]

Automotive Suppliers [19.3]

As shown in this pie chart, automotive interests constitute 40 % of the ESD's 1995 membership, hardly surprising in the world's "Motor City." (ESD)

ESD — The Engineering Society traces its origin to an engineering group founded on June 14, 1895. Out of this group came the Engineering Society of Detroit, which was incorporated on April 15, 1936. Within three months, the membership rose to 1,000. Six years later, the membership reached a total of more than 3,000, and the Society moved into its new headquarters, made possible through gifts from the Horace H. Rackham and Mary A. Rackham fund, administered by the Rackham Engineering Foundation trustees. Today, 100 years from its inception, ESD membership stands at more than 8,000, making it the largest metropolitan engineering society in the world. Listed below are the names of the dedicated leaders who have served as president of ESD down the years.

Past presidents

1902-03 E. E. Haskell
1903-04 F. C. McMath
1904-05 T. H. Hinchman, Jr.
1905-06 J. D. Sanders
1906-07 Benjamin Douglas
1907-08 E. S. Wheeler
1908-09 Francis C. Shenehon
1909-10 William R. Kales
1910-11 T. F. McCrickett
1911-12 Ralph Collamore
1912-13 B. J. Denman
1913-14 E. J. Burdick
1914-15 H. H. Esselstyn
1915-16 Orton W. Albee
1916-17 Horace H. Lane
1917-18 W. P. Putnam
L. R. Hoffman
1918-19 Charles Y. Dixon
1919-20 D. J. Sterrett
1920-21 D. V. Williamson
1921-22 M. W. Taber
1922-23 J. R. McColl
1923-24 L. E. Williams
1924-25 Frank Burton
1925-26 George H. Fenkell
1926-27 George R. Thompson
1927-28 William F. Zabriskie
1928-29 Charles J. Peck

1929-30 John P. Hallihan
1930-31 E. M. Walker
1931-32 A. S. Douglass
1932-33 F. O. Clements
1933-34 H. L. Walton
1934-35 H. S. Ellington
1936-38 John H. Hunt
1938-40 James W. Parker
1941-43 Harvey M. Merker
1943-44 Thomas A. Boyd
1944-45 Elwyn C. Balch
1945-46 Harold S. Ellington
1946-47 Clement J. Freund
1947-48 George R. Thompson
1948-49 S. M. Dean

1949-50 Frank H. Riddle
1950-51 Grant S. Wilcox, Jr.
1951-52 J. L. McLoud
1952-53 Kenneth A. Meade
1953-54 William E. Stirton
1954-55 Earl Bartholomew
1955-56 Raymond Foulkrod
George H. Miehls
1956-57 Earnest Boyce
1957-58 William H. Graves
1958-59 George A. Porter
1959-60 Fred J. Meno II
1960-61 Frederick Bauer
1961-62 Clyde L. Palmer
1962-63 Howard M. Hess

1963-64 Glenn V. Edmonson
1964-65 Walter L. Couse
1965-66 John M. Campbell
1966-67 Donald E. Jahncke
1967-68 Donald N. Frey
1968-69 Harvey A. Wagner
1969-70 Sol King
1970-71 William M. Dull
1971-72 L. B. Bornhauser
1972-73 Morton S. Hilbert
1973-74 James S. Owens
1974-75 Francis E. Cogsdill
1975-76 Robert R. Johnson
1976-77 Kenneth L. Hulsing
1977-78 William P. Panny
1978-79 Jack L. Korb
1979-80 Alfred M. Entenman, Jr.
1980-81 Richard E. Marburger
1981-82 Sydney L. Terry
1982-83 Reinhold M. Tischler
1983-84 Creighton C. Lederer
1984-85 Roy H. Link
1985-86 Alex C. Mair
1986-87 Dr. Serge Gratch
1987-88 Daniel H. Shahan
1988-89 Dr. Stephen R. Davis
1989-90 Douglas R. Allen
1990-91 Edgar E. Parks, P.E.
1992-93 Harry Tauber
1993-94 Lydia B. Lazurenko, M.E.
1994-95 Donald J. Smolenski P.E.

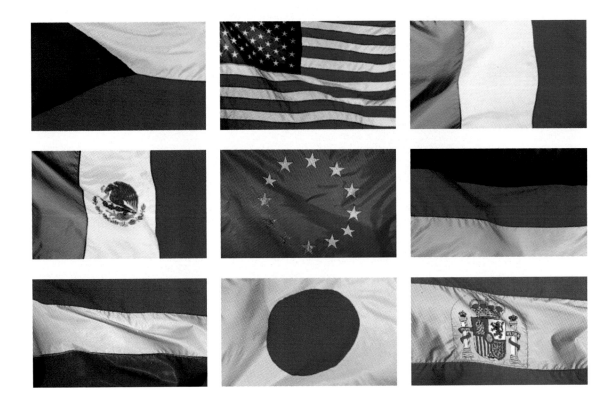

Worldwide. Worldwise.

Leadership in the global automotive market requires both characteristics: Worldwide resources to meet the critical component needs of an industry that produces more than 45 million new vehicles annually; Worldwise savvy to develop, manufacture and market these components in diverse cultural and business environments.

Worldwide. Worldwise. It fits our profile. We're one of the world's leading makers of anti-lock braking systems. And no manufacturer in the world produces a wider range of conventional brake components and assemblies.

We design, engineer and manufacture all of these products at more than 30 locations in 10 countries and on three continents. To know more about a company that's worldwide and worldwise, call (313) 513-5000.

KELSEY-HAYES ™

Where Engineering Excellence Is A Tradition

A Business of
Varity Corporation **VARITY**

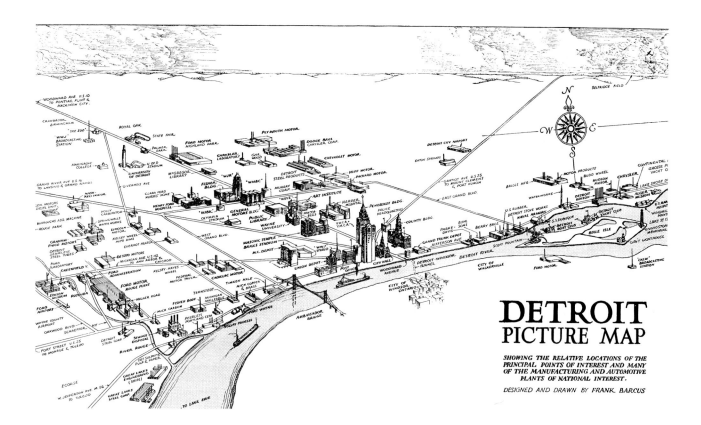

DETROIT PICTURE MAP

SHOWING THE RELATIVE LOCATIONS OF THE
PRINCIPAL POINTS OF INTEREST AND MANY
OF THE MANUFACTURING AND AUTOMOTIVE
PLANTS OF NATIONAL INTEREST.

DESIGNED AND DRAWN BY FRANK BARCUS

Industry & Its Engineers

100 years of technology and innovation

by Dr. Charles K. Hyde

TOP: *This 1939 "birds-eye" view of Detroit shows the location of many companies and installations long forgotten. Through the magic of computer graphics, the publishers were able to correct easily a mistaken identification in the original art. (Bob Cosgrove collection)* **PAGE 7:** *This view of Woodward Avenue at Cadillac Square in the 1880s shows that technology had just barely arrived. Electric light towers rise high over the City Hall lawn (center and right) but the streetcars are still horse-drawn. (Burton Historical Collection, Detroit Public Library)*

I n surveying Detroit's engineering and techno-logical accomplishments, one is tempted to simply reiterate the obvious — Detroit made cars and cars have made Detroit. In the rest of the United States and throughout the world, Detroit is seen as the Motor City and little else. To be sure, Detroit has been the premier industrial city in the world in the twentieth century, the birthplace of mass production and the promoter of mass consumption. American automobile nameplates reflect the contributions of Detroiters to the automobile industry — Ransom E. Olds, Henry Ford, John and Horace Dodge, David Buick, Walter P. Chrysler, and even Antoine de la Mothe Cadillac, the founder of Detroit.

Metropolitan Detroit, however, has also enjoyed a distinguished and noteworthy industrial and technological history well before the coming of the automobile in the early twentieth century. Even after the automobile industry dominated the Detroit economy in the 1910s and 1920s, the city continued to serve as the home for scores of innovative industries other than automobile manufacturing. Today, Detroit is no longer the center of automobile manufacturing, but it nevertheless has remained the intellectual capital of the American automobile industry, with most of the industry's managerial, styling, and engineering talent remaining here.

Many of the articles in this volume provide detailed information about Detroit's highly-diversified industrial base of the 19th century. Some of Detroit's

NO OTHER WEIGHT LOSS PROGRAM HAS BEEN PROVEN AS SAFE AND EFFECTIVE.

If you'd like to see an impressive before-and-after story, just take a look at what steel has done for the modern automobile.

In fact, we've shown how to shed 140 or more unwanted pounds from the body-in-white of a typical North-American four-door family sedan. All without sacrificing safety, *and* at a cost savings of $40 per car. And that's just the beginning. Thirty-one steel producers from

North America, Europe and Asia are now developing steel design concepts for the ultra-lightweight car of the future.

And when you consider steel's recycling track record and affordability, it could very well be the only material that's capable of meeting the government's "Supercar" objectives of high mileage, low cost and increased recyclability.

Which is one thing you can't

say about all the alternative material "fad diets."

Call 1-800-STEEL WORKS for more information about losing weight with automotive steel.

DRIVING THE FUTURE.

American Iron and Steel Institute Automotive Applications Committee

early industries were tied to the rich mineral and forest resources of the Great Lakes basin. Early iron and steel manufacturers located on Detroit's waterfront, on Zug Island, and in Wyandotte; Lake Superior copper interests built a large copper smelter on the Detroit River south of Fort Wayne; and bulk chemical manufacturers located in southwest Detroit and Wyandotte, producing soda ash, salt, and salt products.

KEY INDUSTRIES

The pharmaceutical industry, led by the Parke, Davis Company and Frederick Stearnes and Company, rose to national prominence in the 1870s and remained a major employer in Detroit well into the 20th century. Other key industries included shipbuilding in Detroit, Wyandotte, and Ecorse; the manufacture of railroad cars and wheels; and the stove industry. Detroit was the largest producer of parlor and cooking stoves in the United States in the 1880s and remained a center of stove production until the 1920s. By the turn of the twentieth century, Detroit was a city with hundreds of foundries and machine shops, and thousands of entrepreneurs and skilled workers experienced in the metal trades.

The remarkable growth of the Detroit and Michigan automobile industry in the 20th century and the emergence of the Big Three are recalled elsewhere in this volume. This was largely the story of a few innovative, visionary men — Olds, Ford, Leland, the Dodges, Buick, Durant, Chrysler, and others — who made Michigan the automobile capital of the world. They have achieved a degree of immortality through the

nameplates on their products. Much of the Detroit automobile industry's continued success resulted from hundreds of large and small improvements to the design of cars and to the way automakers built them. Innovations such as the electric self-starter, steel wheels, hydraulic brake systems, high-compression engines, the all-steel body, safety glass, automatic transmissions, and countless other improvements have propelled the auto industry forward.

Unfortunately, the firms which served as suppliers to the automobile companies and the remarkable men who led them are typically relegated to near-obscurity in the history books. The Dodge brothers manufactured engines, transmissions, and other components for Olds and Ford from 1901 until 1914, when they introduced their own nameplate. Other early Detroit suppliers who made substantial contributions to the success of the automobile industry included the Kelsey Wheel Company, later Kelsey-Hayes; the Holley Brothers Company (carburetors); Timken-Detroit Axle Company (roller bearings); the Edward G. Budd Manufacturing Company (steel bodies); and the Fisher Body Company. A list of important suppliers could be expanded to several pages, if space permitted.

NOT JUST CARS

Outside observers of Detroit and even its own residents often view southeast Michigan's economy as completely automobile-oriented. In fact, hundreds of manufacturing firms making non-automotive products have survived and flourished in this century. The Burroughs Adding Machine Company, which moved to

The Team of The Century

As the official publisher for ESD—The Engineering Society, we are proud to present THE TECHNOLOGY CENTURY as a chronicle of ESD's first hundred years of achievements and commitment to industry.

The production of this book was no small task. But with dedication and teamwork, we at Engineerng Technology Publishing, Inc. were able to provide the finest in writing, editing and graphics, as well as a monumental effort to solicit supporting advertisers.

As a contract publisher working integrally with our client and its publications committee, we are expected to offer these turnkey services. And because we have the team of the century, we can.

Introducing our team:

ENGINEERING TECHNOLOGY PUBLISHING, INC.

Barbara Kelvin, *President*
John Kelvin, *Publisher*
David Tell, *Managing Editor, Art Direction*
Kevin Campbell, *Advertising Manager*
Karen Thomas, *Listings Coordinator*
Pat Dewitt, *Advertising Assistant*
Douglas Kelvin, *Pre-press Manager*

ESD PUBLICATIONS COMMITTEE

Roger Bastien, *NARMCO Group*
John T. Benedict, *Standards Consultant*
Mike Davis, *Editor, The Technology Century*
Dr. Ralph Kummler, *Wayne State University*
J. Douglas Mathieson, *ESD Staff*
James Meloche, *French & Rogers*
Rich Moizio, *ESD Staff*
John Petty, *General Dynamics*
Karen Torigian, *Giffels Associates*

Engineering Technology Publishing

27421 Harper Ave. St. Clair Shores, MI 48081
810-774-3530

Detroit from St. Louis in 1904, remained one of the major manufacturers of office equipment and computers until its recent merger with Sperry to form Unisys. Detroit was a significant manufacturer of civilian passenger aircraft with the Stout Metal Airplane Company in the 1920s and the Stinson Aircraft Company (1926-1948). Kelvinator and Frigidaire, both leaders in introducing electric home refrigerators, had early factories in Detroit. Southeast Michigan's diverse products include air rifles (Daisy); toy trains (Lionel), and circuit breakers (Square D), to name a few. More recently, southeast Michigan has become a center for the manufacture of automation equipment, including numerically-controlled tools and robots.

The automobile industry demonstrated its collective manufacturing genius during the Second World War, when Detroit literally became "The Arsenal of Democracy" and produced a flood of war goods. That Detroit could produce large volumes of trucks, tanks, and other wheeled vehicles surprised nobody. But the remarkable speed and flexibility of the auto industry's designers, engineers, managers, and skilled workers in making unprecedented volumes of unfamiliar products astonished the world. Detroit's factories produced aircraft and aircraft engines of all descriptions; artillery; small arms and ammunition; gyroscopes; radar equipment; bombsights; marine equipment; and scores of other products.

Beginning with the Second World War, the auto industry began moving its manufacturing facilities into the nearby suburbs, including Warren, Center Line, Livonia, and Wixom. This migration reflected the shortage of large blocks of vacant land in Detroit by the late 1930s and the desire to build sprawling single-story factories like the Ford Rouge complex. From the 1950s on, the industry built new plants even further afield in southeast Michigan, for example, in Sterling Heights, Lake Orion, Ypsilanti Township, Flat Rock, and Trenton. The auto makers also located their new factories in the Sunbelt and overseas, to further reduce production costs.

PLANT OBSOLESCENCE

Detroit's automakers began to neglect the oldest generation of massive automobile plants starting in the 1960s and finally abandoned and demolished many of them in the 1980s and 1990s. Enormous manufacturing complexes such as Dodge Main, Chrysler Jefferson Avenue, and Uniroyal are gone, while other plants like Ford Highland Park, Packard, and Murray Body Company are used mainly for warehousing. Automobile assembly within the city of Detroit proper continues

only at the General Motors Detroit-Hamtramck Assembly (Poletown) plant and at Chrysler's Jefferson North Assembly plant, both opened in the last decade.

Detroit and southeast Michigan are no longer the center for automobile production in the United States. This region remains, however, the administrative and intellectual capital of the American automobile industry, and in some respects, is the capital of the global auto industry as well. The Big Three have maintained their corporate headquarters here, including the venerable General Motors Building (1922) in Detroit's New Center, designed by Albert Kahn; the Ford Motor Company World Headquarters, better known as "The Glass House" (1956) in Dearborn; and the ultramodern Chrysler World Headquarters (1995) in Auburn Hills.

The Big Three have kept their design, engineering, and testing facilities in southeast Michigan as well. Ford uses the former Ford Airport in Dearborn as a test ground, while its major engineering and design facilities are located in a half-dozen major buildings nearby. General Motors operates a vast proving ground at Milford, dating back to 1923, and an enormous, beautiful Technical Center in Warren, begun in 1956. Chrysler recently opened its new Technology Center (1993) in Auburn Hills, moving its engineering operations from Highland Park. Although the Big Three no longer employ the large factory work force of the past, the auto industry remains a major employer of designers, engineers, technicians, managers, and other white-collar workers.

Later sections of this volume consider Detroit's innovative architects, who designed the auto industry's factories and offices, as well as the expanding city that developed as a result of the industry's success. The importance of Albert Kahn's radically new factory architecture to the success of Henry Ford and the other automakers cannot be emphasized too much. At the same time, we should not ignore the engineers who contributed to Detroit's phenomenal growth in the first half of the 20th century through their work in transportation and public utilities.

The Detroit River, an important conduit for trade from Detroit's founding in 1701, has also been a major transportation barrier until the twentieth century. Detroiters were instrumental in designing and building two tunnels and a bridge to overcome this obstacle. The Michigan Central Railroad Detroit River Tunnel (1909) used steel tubes sunk in trenches for the segment passing directly under the river, a distance of 2,620 feet. This was the first tunnel in the United States to use this particular design. The Ambassador Bridge (1929) had the longest main span (1,850 feet) of any suspension

Detroit's downtown right after World War II was a bustling center of high-rise office buildings and hotels erected just before the Great Depression of the 1930s. Within a few years, urban renewal, expressways and depopulation began to change this appearance dramatically. (The Detroit News)

bridge in the world when completed, but lost its record to the George Washington Bridge in 1931. Finally, the Detroit-Windsor Vehicular Tunnel (1930), which also used a trench-and-tube design, was the third major subaqueous vehicular tunnel built in the United States.

The engineers who designed the various systems which provide southeast Michigan with electricity, gas, water, sewerage, and telephone services have also contributed mightily to the region's growth over the last century. Their work is discussed in detail in several articles in this volume. Space does not permit a listing of all of the men and women who have established southeast Michigan as a center of technological and engineering innovations nationally and internationally, both within the automobile industry and in other areas. The same vitality and innovative spirit is still evident throughout this region and will guarantee that the next 100 years will be southeast Michigan's second Technology Century.

The Engineering Society

*Detroit has
the honor of
many firsts & foremosts*

By Anthony J. Yanik

TOP: *A view of the northwest corner of Warren and John R, the site of The Engineering Society's new building just prior to construction commencing in 1940. (ESD)*

One hundred years ago, 13 young engineers heeded the call of Dr. Mortimer E. Cooley (later to become Dean of Engineering at the University of Michigan), and formed the Detroit Association of Graduate Engineers. Today, as the Engineering Society of Detroit, the Association has grown into the largest local engineering society in the world.

Dr. Cooley's call came at a critical time: he had received word that University Regents were questioning the value of an engineering education, thus opening up the possibility that the engineering program would be dropped or drastically revised.

Realizing that a formal protest from an organized group could be influential, three Michigan engineering alumni, Gardner S. Williams, H. G. Field and H. R. King, drew up a constitution and by-laws for such an organization. These were presented on June 14, 1895, at a dinner in Detroit's Chamber of Commerce cafe. Attending were engineering alumni and most of the University's engineering faculty. They subsequently adopted the constitution and by-laws, and the Association thus was formed with Walter S. Russel chosen as its first president.

The support of the newly formed Association helped dissuade the Regents. They took no action. The College of Engineering has since gone on to earn an enviable reputation for the caliber of its graduates.

MOVED TO DETROIT

In 1901, the Association decided to expand its membership to include engineering graduates of other universities, prompting a name change to that of the "Detroit Engineering Society" (DES). Monthly meetings were moved from Ann Arbor to the St. Claire Hotel on the corner of Randolph and Monroe in downtown Detroit, a location more central to the society's growing membership. A common feature of these meetings was the presentation of a technical paper by one of the members. In time the Society began to sponsor papers by well known authorities working in the engineering field.

After the move to Detroit, the Society resolved to form special committees from among the membership, several of which offered invaluable technical assistance to the city, then grappling with expanding water supply, sewage, and lighting needs. Out of one such committee came a manual of building codes and safety ordinances which was adopted by urban communities in Indiana, Ohio, and southwestern Ontario as well as Michigan.

Soon the little dining room of the St. Claire Hotel became too small for these gatherings, and DES began to meet at various other locations. Residence was taken up in rooms of the Employer's Association in the Stevens Building from 1906 until 1914 when the Society rented offices in a three-story building on 46 West Grand River where it remained for four years.

By this time, DES had expanded from its 13 original members to some 437. This number dropped substantially when many entered the armed services during World War I. The subsequent loss in dues income was instrumental in DES accepting an offer from the Detroit Chamber of Commerce in 1918 to set up its offices in the latter's building at Lafayette and Wayne.

The end of the war did not bring back the expected buildup in membership. In fact, by 1922 attendance at the monthly meetings had become quite meager, indicating a decided drop in interest in a local engineering organization. It was the consensus of those members still active that the lack of a permanent home was at the root of their problem, aggravated by the fact that their identity had become quite lost among all the other activities then taking place within the Board of Commerce building. Numerous meetings were held until members came to the conclusion that they should seek a large, fine home not far from the downtown district, the interior of which they could turn into offices and meeting rooms, and to which they could later add an auditorium.

In 1923, a letter ballot of the members resulted in a 75 percent acceptance for this direction. Within a short time, $15,000 was raised, enough to make a 25 percent down payment on a large, three-story mansion at 478 West Alexandrine, as well as alter its interior. DES officers then proceeded to hire a full-time administration staff to service the monthly meetings, technical sessions and other activities involving the membership. Not long after, they were contacted by the University of Michigan with a request for space within the new quarters where it could conduct accredited extension courses, the first ever to be presented outside the Ann Arbor campus.

HARD HIT

By 1929, with membership having escalated to 871 over the intervening six years, more room once again was needed. The DES Board of Directors voted to seek funds for a major expansion of the Alexandrine property. No sooner had it made this decision than the Great Depression descended upon the nation. By 1934, DES had lost 75 percent of its members because they could not afford to pay their annual dues, minimal though they were. The resulting drain on dues income was so acute that DES no longer was able to meet the tax, mortgage and interest payments on the Alexandrine building.

Faced with bankruptcy, Harold S. Ellington, an architect then DES president, sent a letter to past and present members asking for financial assistance and suggestions by which the Society could regain solvency. One of the recipients of this letter was Bryson Horton, a trustee of the Horace H. Rackham and Mary

Harold S. Ellington (ESD)

A. Rackham Fund. The Rackham Fund had been created in 1933 according to the terms of the will left by Mr. Rackham at his death, and was intended for the support of educational or technical efforts to improve public welfare, education and training. Mr. Horton suggested that DES petition the Trustees for financial aid because of its work in promoting the advancement of the engineering profession. (And the profession did indeed need advancing. In a letter dated March 13, 1934, a Mr. W. P. Thomas mentioned that window washers working for Wayne County were being paid $110 per month, whereas engineers in county employ were earning only $75 per month.)

Mr. Ellington and other DES members held several meetings with the Rackham Trustees beginning in November 1934, meetings that culminated in a grant of about $50,000 which enabled the Society to meet its most pressing financial needs.

Early in 1936, Rackham Trustees approached DES and recommended that it reorganize as a non-profit corporation. Accordingly, on April 15, 1936, DES became the Engineering Society of Detroit (ESD). Offices were rented in the Hotel Statler in downtown Detroit, and the Trustees set aside a $500,000 fund for ESD operations. The fund was to be administered by a newly incorporated Rackham Engineering Foundation.

With more resources at its command, ESD began to expand its operations. Thus the Junior Section of the ESD was formed early in 1937, comprised of members between the ages 21 to 30 who had completed a bachelor of science degree.

Mary A. Rackham (ESD)

RACKHAMS TO THE RESCUE

In May 1937, ESD received a second $500,000 from the Rackham Fund for the expressed purpose of constructing a permanent, new headquarters. Mrs. Rackham then added another $500,000 eight months later from her personal funds.

Indeed a permanent new headquarters was sorely needed. The Society had experienced phenomenal growth over the prior few years, from 523 active members in 1930 to 2,396 in 1938.

Moreover, this growth was characterized by a marked increase in members who were from the automobile industry. For example, in 1914, only 24 represented autos, the most well known being Henry Ford and Howard E. Coffin (one of the founders of the Hud-

OPPOSITE: *At the peak of its activities in the post-war period, the Rackham Building was host to both professional training and social events. This member dance scene, probably in the 1960s, also shows the handsome art deco interior design of the building before a 1974 fire. (ESD)*

son Motor Company). In 1930, this number was only 36 or 7 percent of the active membership. But in 1938, auto company representation swelled to 500.

Once again the University of Michigan entered the ESD picture. After the University announced that it was looking for a suitable building site in Detroit in which it could conduct its extension courses, Rackham interests proposed that ESD and the University of Michigan jointly construct a Horace H. Rackham Memorial Building that would meet both their needs. On Dec. 6, 1938, ESD members voted overwhelmingly in favor of the proposal, and building plans were initiated. As approved by all parties, a single building would be constructed. ESD would own that part in which it conducted its business activities, and U. of M. the remainder including the auditorium, which prompted Rackham Trustees to award U. of M. an extra $1 million for its share of the costs. Final land purchases were made early in 1940 when Mrs. Rackham made a further gift of $750,000. Construction commenced on July 1 of that year, and the splendid Memorial building subsequently was dedicated on Jan. 28, 1942, as the U.S. was entering World War II. Although ESD suspended business activities for the duration of the war, the Rackham Memorial Building remained active as a meeting place for the defense industry and the military.

Once the war ended, ESD resumed its normal operations. To help returning veterans brush up on their engineering training, the Society added review courses leading toward Professional Engineering Registration in the state of Michigan (and continues to offer such courses today).

CENTER OF ACTIVITY

For the next 20 years ESD and the Rackham Memorial Building served as the center of engineering activities within the Detroit area. All Detroit technical societies used it as their meeting place. They were sup-

ported by ESD's publication, "The Foundation," which provided schedules of oncoming meetings and announcements of future technical programs.

In 1953, the Society instituted an annual Outstanding Young Engineers award program. The initial recipient, Donald Frey, justified ESD's choice. He eventually became head of Ford Division, and later chairman of Bell & Howell.

Of critical impact were the riots that occurred in Detroit in 1967 after which the affiliated technical societies gradually began to move their evening technical programs to sites outside of the city. Ultimately their relocation away from the Rackham Building created a substantial drop in ESD revenue from the loss of meet-

ing room rental fees and income from food and beverage sales. Neither loss could be offset by changes in membership dues alone.

Finances became all the more strained in the fall of 1974, when an accidental fire gutted the ESD wing of the Rackham building and created substantial smoke damage to the University of Michigan quarters. The Rackham Building was forced to close for one year while the necessary repairs were made. Included in the repair costs was an extra burden of $1.5 million above insurance claims to bring the rebuilt structure up to an updated Detroit code. On a brighter note, ESD members contributed the funds used to refurnish the interior.

Unfortunately, the affiliated technical societies continued to conduct their evening meetings and conferences away from the Rackham site in the suburbs of Detroit, even after a controlled parking structure was built adjacent to the Rackham building in 1976 at a cost of $1.5 million (of which $1.1 million was underwritten by the Rackham Engineering Foundation).

On the plus side, the ESD wing of the Rackham Building served as a daytime meeting place for several of the Society's 420 corporate members such as Ameritech, Detroit Edison, Albert Kahn Associates and the automobile companies.

REVENUE RELIEF

Nevertheless, the income lost from evening activities continued to be a burden. To compensate, ESD decided to experiment with lectures and technical conferences of its own. It proved to be a wise decision. Today, the return from these two sources alone represents 75 percent of the gross revenues of the Society. In fact the scheduling of ESD lectures and technical conferences became such a demanding effort that the Society opened a Conferences and Expositions Regional Office in Ann Arbor in 1984 to provide the necessary support.

Also worthy of mention is that over the past 20 years, ESD has instituted a number of annual awards that have been favorably received within the professional and local communities. Probably the most significant of these is the prestigious Gold Award annually presented to an engineer, scientist or technician of the metropolitan Detroit area for outstanding work in his or her field. Twenty-three people have been honored thus far beginning with Walker Cisler of Detroit Edison in 1972. Another key event has been the Annual Construction and Design Awards program inaugurated in 1974. Each year ESD selects several outstanding new building complexes completed that year, and honors the members of the construction, engineering and architectural industries responsible for them.

Other awards that have been initiated are the Outstanding Student Engineer Award, first instituted in 1980; the Horace H. Rackham Humanitarian Award, now in its 17th year, which has recognized such luminaries as Elliott "Pete" Estes, Richard Kughn and Lee Iacocca; and the Distinguished Service Award whose 41 past recipients include a former Detroit mayor, Murray D. Van Wagoner.

Special mention also should be made of the Society of Engineers' Wives (SEW) originally formed during the dark Depression days of the 1930s as the women's' auxiliary of the DES. SEW has as its major function the raising of scholarship funds for deserving engineering students from the local area.

More recently, on Oct. 20, 1988, ESD filed an amendment to its Articles of Incorporation changing its tax status from that of a professional and business league to a charitable, scientific and educational organization.

In 1994, a combination of factors developed that forced the ESD administration to make several hard decisions. These factors were a gradual escalation in monthly operating costs (a 350 percent increase over the past 25 years), the need for $8.5 million to bring the property up to the current Detroit fire code and for an additional $2.5 million to initiate changes in building access to meet the Americans with Disabilities Act.

WAYNE 'ESTATE'

The solution came in the form of a proposal from Wayne State University under which the latter would sign a 25-year lease to use the ESD wing of the Rackham facilities and provide the funds necessary to bring the building up to the current fire code. Income from the lease would be sufficient for ESD to meet its share of the Rackham Building's monthly operating expenses. ESD also would continue to occupy office space for its administrative staff.

ESD presently is evaluating whether it would be more beneficial for the staff to move to a location closer to its membership base. A recent demographic study by the Lawrence Technological University for ESD has indicated that the geographical center of the extensive ESD membership is located at 12 Mile and Orchard Lake Roads.

Present needs aside, ESD's record of achievements in promoting the engineering and scientific professions within the Detroit area are recognized across the nation. For generations the society has provided a forum where technical ideas could be discussed and brought to the attention of interested people, ideas that bear on the physical progress of the society in which we live.

Today ESD is one of only three local community engineering societies in existence, the other two being in Chicago and Cleveland. Of the three, ESD is by far the most active, thus the one that most meets the definition of a living engineering society. If the past is any indication of the future, ESD can look forward to the next millennium with utmost confidence that it will provide its members with a leadership that bodes success for the individual as well as the community.

Foundation

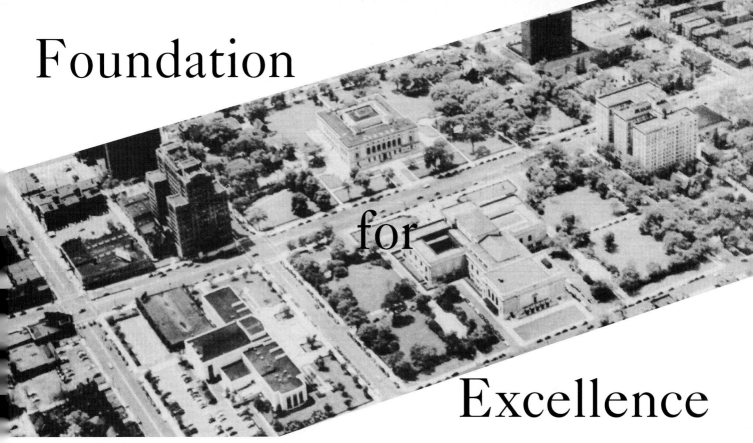

for

Excellence

ESD's philanthropic programs foster progress

by Anthony J. Yanik

A fter more than a decade of evolution and progress, the Engineering and Science Development Foundation, the society's primary fund-raising arm, is launching an effort designed to secure ESD's prominence in educational programming into the future.

FutureLINK, a $5 million campaign, is a collaborative endeavor to bring the International Science and Engineering Fair to Detroit in the year 2000, to endow the Science and Engineering Fair of Metropolitan Detroit in perpetuity, and to create a fund which will guarantee adequate resources for ESD's many educational and outreach efforts.

The Foundation was originally incorporated as the ESD Endowment Fund on July 20, 1984, to bring a sense of focus to the Society's fund- gathering activities. Such focus was lacking as each ESD activity sought its own operating monies, sometimes approaching the same funding sources within the same timeframe, often with disappointing results.

The Endowment Fund office now became the sole fundraising agency for the following groups: The Sci-

ence and Engineering Fair of Metropolitan Detroit, the Engineering and Science Fellowship Fund (which annually supports an engineer or scientist to work in Washington), the Science Court and Research Institute (recognized for its expertise in resolving disputes of a scientific nature), the Engineering Society of Michigan (which sponsors interns for the Michigan Legislative Bureau and provides technical assistance to the state Legislature and other educational programs sponsored by ESD).

Credit for initially molding the Fund into the force it has become were its original officers: Robert Koval (president), Donald Wickham (vice president), and Creighton Lederer, John Harlan, Charles Herdel and Glenn Edmonson (directors).

On Dec. 12, 1984, the Endowment Fund's name was changed to the "ESD Development Fund" to better characterize its activities.

Then, in 1987, The Development Fund introduced what has become its most important fund-gathering activity — the annual Leadership Awards Dinner. Through the Leadership Awards, the ESD honors those individuals who have been outstanding in promoting the advancement of science and engineering and whose

TOP: *An aerial view of the University Cultural Center, taken about 1950, shows how the Rackham Building (lower left) has long been a major structure in the area. The photo preceded major expansions of the Detroit Institute of Arts (center foreground), Detroit Public Library (center) and Wayne State University (rear). The Detroit Historical Museum and the Center for Creative Studies were yet to be built. (The Detroit News)*

Birds flock together for protection. As it turns out, so do we.

Whoever said there is safety in numbers must have been talking about ESD. After all, one of your greatest member benefits is access to valuable insurance coverage. At group rates outsiders envy.

We've not only done all the legwork in finding these plans. We've tailored them to the preferences of you, our members. The coverage is easy to update as your needs change. And it stays with you, even when you change jobs.

To speak with a customer service representative, call 1 800 424-9883, or in Washington, DC (202) 457-6820, between 8:30 a.m. and 5:30 p.m. eastern time. It's one time when the numbers are all in your favor.

 Insurance

Term Life • Disability Income Protection • Comprehensive HealthCare • Excess Major Medical
In-Hospital • High-Limit Accident • Medicare Supplement Insurance

In 1987, The Development Fund introduced what has become its most important fund-gathering activity — the annual Leadership Awards Dinner. Through its Leadership Awards, the ESD honors those individuals who have been outstanding in promoting the advancement of science and engineering and whose community involvement warrants recognition. The dinner takes place annually, and is considered one of the most outstanding events within the professional community. Equally outstanding has been the calibre of the Award recipients over the years:

- 1987 — **F. James McDonald**, president & CEO, General Motors Corporation
  Harold A. Poling, president, Ford Motor Company
  Harold K. Sperlich, president, Chrysler Corporation
- 1988 — **Alfred R. Glancy, III**, chairman & CEO, Michigan Consolidated Gas Co.
  Walter J. McCarthy, Jr., chairman & CEO, Detroit Edison Co.
  S. Kinnie Smith, Jr., president, CMS Energy Corporation & Vice Chairman, Consumers Power.
  William P. Vititoe, president & director, Michigan Bell Telephone Company
- 1989 — **Robert C. Stempel**, president & CEO, General Motors Corporation

- 1990 — **Donald E. Petersen**, president & CEO, Ford Motor Company
- 1991 — **Lee A. Iacocca**, chairman & CEO, Chrysler Corporation
- 1992 — **Lloyd E. Reuss**, president, General Motors Corporation
- 1993 — **Louis R. Ross**, vice chairman & chief technical officer, Ford Motor Company
- 1994 — **Robert J. Eaton**, chairman & CEO, Chrysler Corporation
- 1995 —
  Lee M. Gardner, president & CEO, Masco Tech Inc.
  E. Michael Mutchler, vice president & group executive, GM Power Train.
  Kenneth L. Way, chairman & CEO, Lear Seating Corporation

community involvement warrants recognition. The dinner takes place annually, and is considered one of the most outstanding events within the professional community.

Another name change occurred in 1992 when the Development Fund became the Engineering and Science Development Foundation, now more commonly known as the ESD Foundation.

HONORED AS HOST

In December of 1993, Detroit was honored with its selection as host of the International Science and Engineering Fair in the year 2000, and a new committee was formed to be responsible for its planning, promotion and implementation. To make certain that it would have the financial means of providing this support, the ESD Foundation incorporated this effort into "FutureLINK."

The FutureLINK campaign was launched with a $1 million challenge grant from General Motors in honor of ESD's Centennial celebration. General Electric and Rockwell International also have pledged support of the undertaking which will solicit the support of corporations, foundations, and individuals committed to the enhancement of math, science and engineering education in southeast Michigan.

With the earnings from the Leadership dinners and the returns from the FutureLINK program, it appears that the future of the ESD Foundation is secure. This will guarantee ESD the financial wherewithal to assume a more prominent role in critical educational programming in the future.

The current Foundation's Board members are to be commended for this strong effort: Donald Smolenski, P.E., Ph.D.(president), John G. Petty (vice president), Stanley K. Stines, P.E., Ph.D. (treasurer), Erica Gordon, (secretary), William Birge, P.E., Charles M. Chambers, Ph.D., Susan M. Cischke, Timothy A. Fino, P.E., Norman Gjostein, Ph.D., J. Douglas Mathieson, John Rakolta, Jr., Richard Rossio, P.E., Ronald L. Roudebush, and Paul T. Sgriccia, P.E., (directors).

Conferences, courses, expos & seminars

ESD's Conference and Expositions Office is located at 2350 Green Rd., Suite 190 in Ann Arbor, Michigan. This office researches, develops, designs, and coordinates conferences and expositions for non-members as well as members of the society. ESD has been organizing conferences and expositions since 1971 (seminars and classes were already offered before then) and is continuingly very involved in events regarding information exchange.

- **Advanced Coatings Technologies** addresses the most important concerns of both producers and users of industrial coatings. This currently includes presentations on physical properties of coatings, coatings for plastics, and durability and application.

- **Advanced Composites Conference and Exposition** presentations and exhibits cover body, chassis, powertrain and infrastructure applications, materials science, manufacturing processes and recycling of composites, polymer composites, and metal matrix composites.

- **Air, Water & Waste Technologies Conference and Exposition**, an environmental management conference. This conference and exposition offers a broad range of topics that address the problems and issues of all aspects of environmental management, and provides a forum to bring together engineers, industry and policy makers to discuss the latest in techniques, equipment, regulatory compliance, and management issues.

- **Alternate Fuel Vehicles/Environmental Vehicles Conference and Exposition** provides vehicle manufacturers, researchers, developers, suppliers and regulators of all industries with the opportunity to present their work, products, and perspectives concerning the challenge for improving fuel economy, reducing energy dependency and striving for cleaner air. The ultimate objective of minimal impact on environment and low fuel consumption will require joining the best efforts of all contributing disciplines.

- **Computer Graphics** explores new developments in CAD and CAM. The conferences includes applications of plant floor and engineering office communications and networks such as MAP, TOP, and local area networks in manufacturing. Rapid prototyping and solid modeling are tutorial and workshop topics. Attendees are platform team members and OEM suppliers working on concurrent engineering projects.

- **Construction Activities Committee (CAC)** was established to foster improvements in the construction industry by promoting increased communications and stimulating the exchange of ideas. Bi-monthly business meetings and a luncheon series are coordinated through the committee which is comprised of members representing a balanced cross section of leaders in the construction industry, including design professionals, owners/users, contractors, suppliers and support services.

- **Environmental Response Planning Conference** was developed to assist those in the environmental sector to address the most recent changes in environmental laws affecting ownership of property in Michigan, as well as to evaluate and manage the risks confronting the economic development community and professional service providers. The program has highlighted changes in environmental law; administrative procedures regulating contaminated soil and ground water and the availability of state funding for remediation.

- **IPC Conference and Exposition.** Now in its 24th year, the IPC Conference and Exposition continues to attract an international audience of industry leaders and technical experts. Originally, IPC stood for "International Programmable Controllers." Today, it embraces all of the techniques and technologies affecting plant-floor automation, including controls, sensors, software, communications, systems integration, simultaneous engineering, and training and management of people, as well as programmable logic controllers. In 1995, IPC as the centerpiece joins the **International Robots & Vision Automation Show** and **MotionExpo** and **The Motion Control Applications and Technology Show** to form the single largest event of its kind in North America, **International Automation Week, the Manufacturing Solutions Show**.

- **Industrial Wastewater Treatment Plant Operators Training Course** was designed, with upgrades over the years, to prepare individuals for the state certification exam offered through the Michigan Department of Natural Resources.

- **Landfill Design/Geosynthetics Conference** will focus on cold climate landflll design, with emphasis on liner frost damage and protective measures needed. Leachate treatment and management and gas generation will be the focus of special tutorial sessions. This conference is designed for environmental engineers, consultants, contractors, owners and operators.

- **Laser Applications/Optics Conference & Exposition** provides a forum for the introduction to lasers and electro-optics in the manufacturing arena. In addition to current applications, new and proposed systems are discussed.

- **Michigan Industrial Expo (MIE)** is the marketplace where plant managers, owners, engineers, manufacturing and purchasing personnel will view industrial applications, products, and services. MIE will facilitate business in Michigan and the Great Lakes Region bringing users to the people who manufacture and provide business services here.

- **OSHA Courses** are a series of classes presented to train people who deal with, or may potentially deal with, hazardous materials and substances. Courses are designed to fulfill OSHA requirements for initial, update, and supervisors' training.

- **Professional Engineers Licensing Review Course** prepares attendees for the State of Michigan P.E. licensing exam.

- **Society for Environmental Sciences (SES)** is a group of environmental professionals whose mission is to promote the free exchange of ideas and technologies; communications between government, business, academia citizens and activists having a stake or concern in environmental issues; and the study of environmentai science and engineering among today's youth. Business meetings and a luncheon series promote this mission to members and non-members

The University of Michigan
College of Engineering

GRADUATE PROFESSIONAL PROGRAMS

- **Master of Engineering**

  Aerospace Engineering

  Air Quality

  Applied Remote Sensing and Geoinformation Systems

  Automotive Engineering

  Concurrent Marine Design

  Construction Engineering and Management

  Manufacturing

  Occupational Ergonomics

  Optical Engineering and Ultrafast Technology

  Radiological Health Engineering

  Space Systems

- **Master of Science**

  Technical Information Design and Management

Qualified candidates at participating work sites may pursue graduate degree programs through the Michigan Engineering Television Network (METN). Off-campus students meet the same rigorous standards and earn the same academic credit as on-campus students. For information on METN programs, contact METN at (313) 936-0252.

FOR INFORMATION

Graduate Professional Programs
University of Michigan
College of Engineering
2423 EECS Building
Ann Arbor, Michigan 48109-2116 U.S.A.
Tel: (313) 763-2304   Fax: (313) 763-9487
e-mail: frbec@engin.umich.edu

Engineering Education

Michigan has led in establishing sites, advancing standards for technical training

by Dr. Richard E. Marburger
and Bruce J. Annett Jr.

TOP: Built in 1908 as an orphanage, this building in Highland Park later was the first home of the Ford Trade School. Later, in 1932, Russell Lawrence leased the facility from Ford to house the new Lawrence Institute of Technology. What is now Lawrence Technological University remained in the building until 1955, when it moved to a new campus in Southfield. (LTU)

hile the application of engineering principles dates back many centuries, the advent of profession of engineering is relatively recent. Until the mid-1700s, the title of "engineer" was reserved for those who constructed engines of war or built works designed to serve military purposes, such as fortifications.

At about the time of the American Revolution, "civil engineer" came into use, describing those occupied with building roads, canals, harbors, docks, drainage projects, and machinery of all types. Military or civil would remain the only two classifications of engineering well into the industrial revolution. About 1850, the field of mechanical engineering was recognized, followed shortly after by mining engineering. As the 19th century closed, engineering began an era of enormous expansion and specialization that continues today.

No Formal Training

At the time of the Civil War, most individuals calling themselves engineers actually had no formal training. Instead, they had learned through apprenticeships. Engineering as an organized profession had its origins among surveyors, particularly those employed in locating railroad routes.

As the great wave of settlement enveloped Michigan through the 1870s and 1880s, the number of surveyors increased steadily in response to the burgeoning

22

Wayne State University

WHERE THE FUTURE STARTS:
WAYNE STATE UNIVERSITY COLLEGE OF ENGINEERING

Chemical
& Materials Science
Civil & Environmental
Electrical & Computer
Industrial & Manufacturing
Mechanical

Wayne State University, with the largest College of Engineering in Metropolitan Detroit, has day and evening classes to suit your needs.

We will provide you with a quality education because our programs have all the rewards and responsibilities of the engineering profession itself.

COLLEGE OF ENGINEERING

For our brochure and video call (313) 577-3780, or write to:
College of Engineering
Student Affairs Dept.
Wayne State University
Detroit, MI 48202

enrolled in engineering!

Recognizing the development of Michigan's vast reserves of iron ore and copper and the need for personnel with the necessary mining skills and technical knowledge, what is today Michigan Technological University was founded in 1885 at the Upper Peninsula town of Houghton.

Southeastern Michigan's growth as a manufacturing center, particularly for such goods as cooking and heating stoves, carriages, the new-fangled automobile, and other mass-produced items at the dawn of the 20th century helped lead to the launch of the state's first independent engineering college in 1891 by the Young Men's Christian Association (YMCA) — the Detroit Institute of Technology. For the same reasons, in 1911 what is today the University of Detroit-Mercy began a college of engineering. In 1928, the Detroit Board of Education established a school of engineering which in 1934 became part of what is now Wayne State University. What is today Lawrence Technological University began offering engineering degree programs in 1932, later adding architecture and engineering technology programs.

The major American automobile companies further influenced technological education in Michigan in the early part of the 20th century by establishing educational divisions. In 1919, General Motors Corp. set up what eventually evolved into today's independent GMI Engineering and Management Institute. Its first engineering degree was offered in 1923. In 1931, Chrysler Corp. launched the Chrysler Institute of Engineering which for many years offered degrees up to the master's level for employees. In 1916, the Henry Ford Trade School was introduced by the Ford Motor Co. The top trade school grads were admitted to the Ford Apprentice School where they were able to take coursework related to automotive design and manufacturing.

The University of Michigan-Dearborn began offering engineering degrees in 1960, Western Michigan in 1961, Oakland University in 1963, and Saginaw Valley in 1984. Calvin College launched engineering degree

need. The scope of engineering interests and activities also widened. It was a period of unprecedented infrastructure expansion, invention, harnessing of new technologies, and manufacturing for newly developing mass markets.

Technological education in the United States has its roots in the military training offered at West Point. Engineering education formally can be traced to the 1820s, when Norwich University in Vermont and Rensselaer Polytechnic Institute in New York were founded, the latter with the goal of "applications of science to the common purposes of life." A naval engineering program was begun at Annapolis in 1845. In 1850, Rensselaer became a four-year engineering college.

Founded in 1817, the University of Michigan became, in 1853, the fourth collegiate institution in the nation (and the first in the state) to offer courses in engineering. It was sixth in the nation to confer degrees — the first two students were graduated in 1860.

In 1882, University of Michigan professor Charles Greene proposed a bill to the Legislature aimed at eliminating incompetent surveyors and improving property records — one of the first such laws in the country. County surveyors strongly opposed the measure, and it took until 1919 to adopt the first Michigan law requiring registration of surveyors, engineers, and architects. In 1937, a new act created the State Board of Registration for Architects, Professional Engineers, and Land Surveyors.

The state's second collegiate program in engineering began in 1885 at what is today Michigan State University. By 1896, one-third of MSU's students were

OPPOSITE: *Ransom E. Olds donated funds to build Michigan State University's former Hall of Engineering during World War I after an earlier structure burned to the ground. At the time, State was still known as Michigan Agricultural College. (Michigan State University Archives and Historical Collections)* ABOVE: *University of Detroit-Mercy's Engineering & Science Building, a facility for the education of thousands of auto industry engineers since it was founded in 1911. (University of Detroit-Mercy)* BELOW: *The University of Michigan's earliest engineering lab, shown here, was established in 1881. A much larger building, also long gone, replaced this tiny structure in 1897. (U-M College of Engineering)*

programs in 1985. Ferris, Central Michigan, and Lake Superior State Universities are among major providers of engineering technology degree programs. Many of Michigan's leading community colleges have also developed strong associate programs in technological fields.

Clearly, there have been huge advances within the engineering profession in Michigan over the past 100 years. Most of this advancement is tied directly to improving collegiate education, developing standards of performance, and being relentlessly dedicated to quality.

Michigan has evolved as a world leader in technology and its applications, as well as being a center of manufacturing, research, and development. If history is any measure, it is expected that the strong partnership between industry and the state's collegiate institutions devoted to engineering and related technological professions will continue to prosper and thrive.

THE WISDOM OF EXPERIENCE.
THE VISION OF YOUTH.

Since the founding of Masco in 1929 on a modest contract to supply tube nuts to the Hudson Motor Car Company, The Masco Family of Companies has virtually grown-up with the ESD. As we greet the challenges of the next century of engineering and manufacturing together, we do so with the wisdom of our experience, combined with the vision of our youth. We look forward to the advances in science and engineering that await us in the next one hundred years.

Masco Corporation, MascoTech, Inc. and TriMas Corporation salute the one hundred year anniversary of the Engineering Society of Detroit.

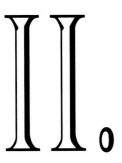

II. $_0$ Auto Motivated

Inventors, investors, industrialists
remake Detroit in a new image

America's first successful motorcar was the Duryea, shown here after it had won the first auto race in which the contestants finished, run in Chicago on Thanksgiving Day 1895. J. Frank Duryea is at the tiller. (National Automotive History Collection, Detroit Public Library)

A Vehicle for Progress

The Auto Industry 1895-1995

by Anthony J. Yanik

The Duryea brothers had been tinkering with their gasoline powered vehicle for several years. In 1895, they decided that it was time to build others like it to sell. Thus they incorporated the first company to manufacture gasoline cars in the United States. By the close of 1896, they had managed to build 13 of them.

Soon they were joined by other more successful early gasoline engine proponents such as Elwood Haynes, the brothers Apperson, Alexander Winton and Ransom E. Olds. Competing against these gentlemen were the powerful motorcars, and the twin brothers, Francis and Freeland Stanley, who had popularized the steam car. Within the decade following the turn of the century, however, it became apparent that neither electrics nor steamers could offset the advantages of gasoline power. By 1905, the latter dominated sales. Focus of that domination was the Detroit-Lansing area from which came two of every three automobiles produced in the nation. Cadillac, Ford, and Olds alone were responsible for half that number.

The man most instrumental in forming this early Michigan-based dominance was Ransom E. Olds. His low-priced runabout had become so popular that he found himself experimenting with the rudimentary techniques of mass production in order to keep up with sales. Olds' output rose from 425 cars in 1901 to 6,500 in 1905 — almost twice that of Cadillac, its nearest competitor. The successes of Olds, Cadillac and

Delphi Interior & Lighting.
Helping you make it.

By making it simple for you. The world is continually looking for a higher quality, more technologically advanced vehicle. A vehicle that offers better value.

Our expertise in lighting and interiors can make it easier for you to meet that goal – through superior global resources and our dedicated staff of designers and engineers.

We'll manufacture a complete lighting or interior system for you. Or we'll build modules or individual components and manage their integration into your vehicle. The result: making it simpler for you to build a better looking, better functioning vehicle. A vehicle that sells. And in this business, that is the bottom line.

So team up with Delphi Interior & Lighting Systems. We'll help you make it.

DELPHI
Interior & Lighting Systems
The power to simplify.

For more information, please call 1-800-788-8758.

Sales Headquarters: Warren, Michigan · Rüsselsheim, Germany · Akishima, Japan

Since 1919,
we've covered some of the best cars in the world...

And we still do.

Back when our name was Rinshed-Mason, our varnish protected the Model T's wooden frame. Since then, both our name and products have evolved.

Today, as BASF, we serve automakers from research facilities and plants throughout the world. Many cars built in North America are finished with our coatings. But that's just the

beginning. Now we're ready to join automakers in developing coatings for the 21st century.

Call us at 1-800-949-2273. We'll help you prepare for the future.

BASF Corporation

Automotive OEM Coatings

Horse vs. Buggy

Early cost comparison favors the horseless carriage

courtesy of Chris Dyrda
(Reprinted from *Scientific American*, Jan. 27, 1900)

SCIENTIFIC AMERICAN

[Entered at the Post Office of New York, N. Y., as Second Class Matter.   Copyright, 1900, by Munn & Co.]

A WEEKLY JOURNAL OF PRACTICAL INFORMATION, ART, SCIENCE, MECHANICS, CHEMISTRY, AND MANUFACTURES.

Vol. LXXXII.—No. 4.
Established 1845.   NEW YORK, JANUARY 27, 1900.   [$3.00 A YEAR.
                                                                                       WEEKLY.

A village resident in one of the English counties has communicated to a local journal an estimate of the relative cost of keeping an automobile and horse and carriage. He arrives at an economy in favor of the motor of $47.75 on the total expenses for the year, and he does it thuswise: The cost of the horse is $115, and of the dog-cart $135; the interest on which outlay, at 4 1/2 per cent for one year, is $11.25; the keep of the horse, at $2.50 a week (it must be remembered that these prices are for keep in a country village), and license and shoeing, bring up the total expense for the year to $159. This he compares with a five-horsepower automobile costing $850, the interest on which, at 4 1/2 per cent for the year, is $38.25. Adding to this a tax of $21 and expense of $52 for fuel (petrol in this case), at the rate of 75 cents for 35 miles, and 25 cents for the same distance for lubrication, he reaches a total annual expense of $111.25.

It will be noticed that in the above estimate there is no repairs account, an item which we think the average unskilled automobilist of the future will find to be, perhaps, the most serious of all, outside of fuel. In this case, however, the automobilist was something of a mechanic, possessing a lathe, a vise, etc., and he was equal to making all ordinary repairs himself; moreover, he argues that in any case the accidents that may happen to a horse, and the more or less frequent visits of the veterinary, will fairly well offset repairs to the automobile.

Just here we would suggest that in view of the fact that the mechanism of the automobile is necessarily complex, and in many forms of motor susceptible to easy disarrangement, it would be well for all intending purchasers to acquire some elementary knowledge of the simpler tools of the mechanic; and we think it is not unlikely that the coming rage for automobilism, which, unlike that for the bicycle, will prove to be lasting, will give an added impetus to the study of practical mechanics in our schools and colleges. In any case the business of automobile repairing will be one of the most important and profitable of the new industries of the future.

later Ford were all the more notable because they survived the lively scramble for sales that ensued during the first decade of the 20th century as some companies vied for leadership.

Nor was this struggle for eminence limited only to automobiles. About 444 truck manufacturers had attempted to enter that market between 1901 and 1910. Only a few endured, among which were Mack, White, International Harvester, and the Grabowsky brothers' Rapid Motor Vehicle Company later purchased by Will Durant. Rapid then was combined with the Reliance Motor Truck Company to form the General Motors

Truck Company, now called General Motors Truck and Bus.

In 1908, two events took place that profoundly affected the U.S. auto industry: Henry Ford introduced his everyman's car, the Model T, and William Durant formed General Motors (GM), a conglomerate of independent automobile companies.

ACCELERATED PRODUCTION

Ford's emergence as an automobile power was the swifter. The low priced Model T, introduced in 1908, soon came into such great demand that Ford moved into a large new factory. In 1913, he introduced the elements

Henry M. Leland, shown here in his office at Lincoln Motor Car Company about 1920, was a pioneer precision parts builder for the infant auto industry in Detroit, later making a success of Cadillac Motor Car Company with interchangeable parts before it was folded into General Motors. (National Automotive History Collection, Detroit Public Library)

of the moving assembly line to further increase production. This extraordinary change in the techniques of manufacturing allowed him to produce 146 Model Ts per hour in 1913. By comparison, in 1908, initial Model T production at best was only 7.5 per hour.

General Motors developed more slowly. Prior to World War I it received stiff competition from such significant new marques as Dodge, Hudson, Hupmobile, Studebaker, Nash, Maxwell and Willys-Overland, if the likes of Ford was not enough. Willys-Overland proved to be a particularly stubborn competitor, running second in sales to Ford for most of the World War I decade. Willys then overextended itself financially and had to be rescued by the Chase National Bank. Willys made a strong recovery in the late 1920s only to be hard hit again by the Great Depression.

When Alfred E. Sloan took over the leadership of GM in 1923, that corporation's fortunes began to change dramatically. In less than 10 years, Sloan brought Ford's 16 year dominance as industry sales leader to an end.

A significant new entry within the automotive circles was Chrysler Corporation, organized in 1925. Chrysler moved quickly into upper sales ranks after it

purchased Dodge Brothers in 1928. (Dodge was a larger company than Chrysler at the time.) The following year Chrysler captured over 8 percent of the sales market, a notable achievement at a time when two of every three cars sold in the U.S. were either a Ford or GM product. The Big Two had now become the Big Three.

AUTOMAKER ATTRITION

Competition was further reduced during the Great Depression which put to rest scores of venerable, low volume car companies that had existed for over 20 years, such as Cunningham, Auburn, Franklin, Locomobile, Marmon, Peerless, Pierce-Arrow, and Stutz. By 1935, ninety percent of cars sold across the nation came from the Big Three. The remaining ten percent was divided among other Depression survivors — Hudson, Packard, Nash, Willys-Overland, Graham Paige, and Studebaker.

The '30s also were notable for another far reaching event — the signing of a labor agreement between the United Automobile Workers and General Motors on February 11, 1937. This agreement led to the eventual unionization of all automobile workers.

Immediately following World War II, the industry settled into a Big Three and Little Four (Hudson, Nash, Studebaker, Packard) relationship with one new builder, Kaiser-Frazer, pushing for entry. Introduced in 1946, Kaiser-Frazer sales leaped to 4.8 percent of the market by 1948 during the post war car sales boom, but it could not sustain that pace. Eight short years later it dropped out of the passenger car scene.

The Little Four also did very well immediately following World War II, then ran into troubled times. In May 1954 Hudson and Nash merged, followed by a name change to American Motors Corporation (AMC) in 1957. AMC had success during the 1950s and 1960s because of its focus on smaller size vehicles. It was absorbed by Chrysler on Aug. 5, 1987.

Studebaker and Packard also merged in 1954. However, the hallowed Packard nameplate survived only four more years. Studebaker cars continued to be built until 1966, then too passed from the scene.

Undaunted by the problems of the independents, Volkswagen began exporting a limited number of its compact Beetle cars into the U.S. in the 1950s. The Beetle soon caught the public imagination. By 1959, its sales had grown to 668,070 units and U.S. manufacturers countered with a flurry of compact car models putting a temporary halt to the import surge.

JAPANESE 'INVASION'

Such respite was short-lived. By 1968, Volkswagen

When this photograph of the General Motors Corporation board of directors was taken May 15, 1928, they were the most successful such group of men in the world. GM had overtaken Ford for the Number One sales position— a distinction still held 67 years later — developed management concepts later to be widely copied and introduced annual styling changes, "stair step" marketing of automotive products and numerous technical advances. The three men in the center of the front row are Alfred P. Sloan, Jr. (GM president 1923-37), Charles E. Wilson (president 1941-53) and William S. Knudsen (president 1937-1940). General Motors) **BELOW:** William Crapo Durant, the former Flint buggy manufacturer whose organizational zeal led in 1908 to the creation of General Motors Corporation, in 1911 to the Chevrolet Motor Company and finally his ill-fated Durant Motors in 1921. He was president of GM from 1916 to 1920. (General Motors)

sales had regained their momentum and were up to one million units. Such numbers enticed Japanese car makers to enter the U.S. automotive market with small cars of their own: Nissan in 1965, Toyota in 1966, Mazda in 1972, and Honda in 1974.

Adding to the burden of problems represented by such new competition was the intervention of the federal government in 1966 when President Lyndon Johnson signed the National Traffic and Motor Vehicle Safety Act. This Act regulated the safety performance of vehicles sold within the United States. Other federal interventions were the Clean Air Act of 1970

regulating tail pipe emissions, and the Energy Policy and Conservation Act of 1980 mandating a series of gradual increases in fuel economy.

In October 1973, the industry was rocked by the Arab oil embargo, and in 1979 by Iran hostage crisis. Gas prices jumped to $1.30 per gallon, causing U.S. car sales, especially of full-size models, to fall to their lowest point in 25 years. Market penetration of the smaller-sized Japanese cars benefited from the crises, rising from 10 percent in 1978 to a peak of 35.6 percent in 1989. This number has since dropped as U.S. makers countered with products especially focused on quality and durability, and reduced the costs of their manufacturing operations. These actions should stand them in good stead as they head into the next millennium.

Looking back over the last 100 years of automobile manufacture, it is difficult to imagine that it represents the entry (and in most cases departure) of well over 3,000 makes of cars and trucks. More importantly, it has witnessed the entry of countless, colorful, daring entrepreneurs, each of whom has contributed in some way to the success of the most valued industry, and thereby to the success of the United States as a world industrial power.

The most exciting material we work with.

It's you. Has been for nearly a century now. There are reasons
for that. You create and execute designs and concepts
that have never been before. Your customers demand them.

You expect certain things from us. You expect more than
innovative materials and application assistance. You
expect solutions. Solutions that save time and exceed
previous boundaries, that surpass environmental
expectations. Solutions that improve value and costs
continuously. Solutions that can give you a competitive lead
of two years or more.

We are DuPont Automotive. We've been with you since the
ragtop era, the first fast-drying finish and the first engineering
plastic. We're as new to you as a completely plastic radiator,
environmental resistant finishes and economically recyclable
composites. And you? You are everyone in the automotive
industry... every engineer, stylist, purchaser and planner who
says "I wonder if..."

**More Technologies
For Better Solutions...
Start With DuPont.**

Together, we make the probable possible. Together, we can
make an exciting, material difference in the future of the
industry. For more information on automotive materials,
call 1-800-533-1313.

DuPont Automotive

In 1924, General Motors pioneered construction of a private proving ground for testing prototype models. At the time of this 1935 photo at GM's Milford property 40 miles northwest of Detroit, GM engineers were racking up 21,000 test miles daily over the 1,268-acre's site's 23 miles of roadways. Shown are a Buick sedan and a Chevrolet coupe. (General Motors)

The Massive Manufacturer

GM broke the mold in its birth, growth as corporate giant

The founding of General Motors on Sept. 16, 1908, drew little attention. Motorcar firms were appearing virtually everywhere. Success for the young automotive concern was not predestined. There was no guarantee of a place in the market or assurance of any profit. Of the nearly 1,000 companies that tried to build and sell motor vehicles prior to 1927, less than 200 continued in business long enough to even offer a commercially suitable vehicle.

Most of the companies that comprised the fledgling General Motors Company were weak, and their operations were uncoordinated. Many were in debt. It was not until the 1920s, when a new concept of management was forged and a new concept of product emerged, that GM really began to prosper.

General Motors' sales for its first full fiscal year ending Sept. 31, 1909, totaled 25,000 cars and trucks, 19 percent of total U.S. sales. Net sales totaled $29 million and its payroll at the peak of the manufacturing season numbered more than 14,000, mostly in Michigan.

Today, General Motors is a global company with almost 300 operations in 37 states and 171 cities in the United States. In addition, GM of Canada operates 20 plants, GM de Mexico operates 5 locations, and GM has assembly, manufacturing, distribution or warehousing operations in 51 other countries, including equity interests in associated companies.

In 1994, GM sold 8,328,000 cars and trucks worldwide with net sales of $155 billion and employment averaging 693,000 workers.

GLOBAL CONGLOMERATE

GM's global business operations are in six sectors: North American Operations and International Operations, which together comprise the largest worldwide manufacturer of cars and trucks; Delphi Automotive Systems, the world's largest and most diversified automotive supplier company; General Motors Acceptance Corporation, providing automotive financing and other financial, insurance and mortgage services; Electronic Data Services, global leader in application of information technology, and Hughes Electronics Corporation, a leader in automotive electronics, space and defense electronics, and telecommunications technologies.

The nucleus of the original General Motors was the Buick Motor Car Company. It was formed in 1902 by David Buick in Detroit and later moved to Flint, Michigan, where William Crapo Durant, "king of the carriage makers," took control. Durant, who brashly predicted that "a million cars a year would someday be in demand," oversaw Buick's rise to become the second largest and most influential automobile manufacturer in the country. He also began organizing a network of suppliers and producers.

When General Motors Company was incorporated as a New Jersey firm, Flint had a population of about 25,000 and four streetcars. It was more than three months before Flint papers carried a single story about the new enterprise.

Early members of the infant GM family of cars were Buick, Oldsmobile, Cadillac, Oakland (now Pontiac), Ewing, Marquette, Welch, Scripps-Booth, Sheridan and Elmore, together with Rapid and Reliance trucks.

Only four of the car lines — Buick, Oldsmobile, Cadillac and Oakland — continued making cars for more than a short time after their acquisition by GM. Chevrolet became part of the corporation in 1918.

By 1920, more than 30 companies had been acquired by General Motors, by purchase of all or part of their stock. Two were forerunners of major GM subsidiaries — the McLaughlin Motor Company of Canada (which later became General Motors of Canada Limited) and the Fisher Body Company.

Although legally a New Jersey corporation, all of GM's original facilities were in Michigan, and Durant encouraged other firms to locate their facilities in the state.

'SELF STARTER'

About the same time GM was getting started in Michigan, an engineering development that was to become influential in GM's subsequent leadership in research was occurring in Dayton, Ohio — the introduction of the electric self-starter. Designed by Charles F. "Boss" Kettering at his Dayton Engineering Laboratories Company, it first appeared on 1912 Cadillacs.

By doing away with the dangerous and unpredictable hand crank, it definitely popularized motoring and made motor cars accessible to a greater part of the population.

"Boss" Kettering later became the technical mastermind of General Motors, in charge of its unparalleled research and engineering programs. He joined GM in 1920 when his company was merged into GM, and moved the Research Laboratories to Detroit in 1925. He remained with GM until his retirement June 2, 1947.

The company officially became General Motors Corporation on October 13, 1916, when incorporation papers were filed in Delaware.

During World War I, GM, turned its facilities and experience to the production of war materials.

Despite having no experience in manufacturing military hardware, the young American automobile industry completed a turnaround from civilian to war production within 18 months. The result was an outpouring of weaponry credited with the winning of the war, changing the face of Europe, and launching the United States into the role of a world power.

It was also in this same period that Alfred P. Sloan, Jr., first became associated with Durant. Sloan was head of the Hyatt Roller Bearing Company of Harrison, New Jersey, when Durant brought it into General Motors through acquisition of the United Motors Corporation. He went on to guide General Motors from 1923 until 1956, first as president and then chairman.

Under Sloan's direction, General Motors grew from a firm that accounted for about 10 percent of U.S. new car sales in 1923 to become the largest producer of cars and trucks in the world.

"Billy" Durant created an enterprise that in 1908 consisted of just one truly successful auto manufacturer (Buick), but it contained the building blocks to become a multifaceted corporation. His entrepreneurial creation was later directed by men with the abilities to harness and organize its potential during an expansionary period of U.S. industry — when the U.S. and General Motors were entering the era of modern management.

Tottering Titan

By 1920, in the midst of a nationwide economic crisis, GM was on the verge of financial collapse. The crisis marked a turning point in General Motors' history. New men were asked to assume its leadership. A new concept of management was forged and a new concept of product emerged: coordinated policy control replaced the undirected efforts of the previous years.

As its principal architect, Sloan was credited with creating not only an organization which saved General Motors, but a new management concept that was adopted by many of the world's large, diversified businesses.

Fundamentally, the concept involves coordination of the enterprise under top management, direction of policy through top-level committees and delegation of operating responsibility throughout the organization. Within this framework, management staffs conduct analysis, advise policy committees and coordinate administration. General Motors thus became an organization of organizations.

Sloan's idea was to establish "decentralized operations and responsibilities with coordinated control." At the individual level, his policy was simple: "Give a man a clear-cut job and let him do it."

His new product concept was "a car for every purse and purpose." It led to different kinds of vehicles for different customers, and continuing improvement of all GM vehicles. People were buying more than just basic transportation. They also wanted comfort, good looks, performance that was better than just adequate, and above all, periodic improvement.

OPPOSITE: *The first Chevrolet produced in the company's Detroit plant in 1912. Founder Billy Durant is standing in the bowler hat at the right and his son Cliff is at the wheel. Namesake race driver and car developer Louis Chevrolet is standing hatless. The following year Chevrolet moved to Flint. It was merged with GM in 1918. (General Motors) RIGHT: From the 1920s to the 1980s, GM prided itself on its engineering advances and styling leadership. Shown here in 1958 are the executives responsible for that reputation in the peak years of the 1950s. Left to right are Charles A. Chayne, vice president in charge of Engineering Staff; GM president Harlow H. Curtice; Dr. Lawrence R. Hafstad, vice president in charge of Research Staff, and Harley J. Earl, vice president in charge of Styling Staff. (General Motors)*

Some of our capabilities:

Simultaneous Engineering

Processing

Prototype Tools & Dies

Quality Control & Inspection

Production Tools

Production Dies

Automated Welding Systems

Hot Test & Dynamometer Systems

Laser Applications

All of our commitment:

Quality

Progressive Tool & Industries Co. pledges to support, promote and participate in a quality improvement process and to provide our staff with the education, resources and environment necessary to supply defect-free products and services in accordance with your requirements.

PICO

THE MARK OF QUALITY®

Progressive Tool and Industries Co.

21000 Telegraph Road • Southfield Michigan 48034 • (313) 353-8888

GM's 50th anniversary models in 1958 were the new Chevrolet Impala (front), with Pontiac, Oldsmobile, Buick and Cadillac parked behind. (General Motors) OVERLEAF: In the 1970s, the industry was battered by the effects of the oil embargo, which doubled energy costs and caused buyers to seek out economy cars — which were almost all imports. GM's answer was this 1976 Chevrolet Chevette, produced both in North America and overseas. (General Motors)

GM developed many automotive "firsts" which enhanced its success. Prior to World War II, they included the first all-steel one-piece roof, two-cycle diesel truck engines, independent front-wheel suspension and automatic transmission.

By 1941, GM accounted for 44 percent of total U.S. sales, compared with 12 percent in 1921.

Even before America's entry into World War II, GM was turning out weapons along with automobiles. After the U.S. went to war in 1941, GM's industrial skills were totally applied with great effectiveness. Civilian automobile production was halted early in 1942 and the Corporation's plants were completely converted to the war effort.

GM's contributions during World War II are a dramatic example of the vital importance of a nation at war being able to call upon well-managed and experienced industrial resources. From 1940 to 1945, GM delivered defense material valued at $12.3 billion.

WARTIME CHANGEOVER

The success of GM's tremendous wartime role lay in its management philosophy. Decentralized, highly flexible operations with localized management capabilities made possible the total and almost overnight conversion from civilian production to war material capability, on a timetable never believed possible by the enemy.

GM's war production spanned products from the tiniest ball bearings to massive tanks, naval ships, fighter planes, bombers, guns, cannons, and projectiles. GM alone turned out 13,000 airplanes and one-fourth of all U.S. aircraft engines.

Car-making resumed after the war, and postwar economic expansion saw production and sales soar. The cars of the '50s were all-new, and their styling captured the public's pent-up desire for change.

For GM in the United States, the decade of the '50s was one of celebrations, sales records, anniversaries, and ingenious innovations in styling and engineering. Cadillac celebrated its 50th anniversary in 1952; the following year in June, Buick built its seven-millionth automobile. GM's 50-millionth automobile, a 1955 Chevrolet Bel Air, rolled off an assembly line in November 1954, and the Corporation celebrated its 50th anniversary in 1958 with a year-long Golden Milestone celebration. Cadillac built its two-millionth car the same year — just eight years after reaching its first million mark.

Through the 1960s GM set more sales and earnings records, and added several new cars and trucks to its model ranges to serve more diverse needs and tastes of customers. New small cars were introduced by Chevrolet, Buick, Oldsmobile and Pontiac and new front-wheel drive luxury cars were launched by Oldsmobile and Cadillac.

General Motors was also growing as a worldwide enterprise during this period. The company's business had reached beyond the U.S. almost from its beginnings; exports of American-built vehicles began in 1911, only three years after the company was formed. In the 1920s, GM opened plants to assemble some of GM's American cars in Europe, South America, Africa and the Far East; there were 19 by 1928.

But the major European markets outside the U.S. and Canada called for different kinds of cars, and GM determined it needed to compete as a European manufacturer. Accordingly, GM purchased Vauxhall Motors of England in 1925, and Adam Opel of Germany in 1929.

POSTWAR EMPIRE

Following World War II, General Motors revived Opel and Vauxhall and expanded its worldwide operations. Full engineering and manufacturing operations were also established in Brazil and Australia, and assembly and marketing operations were expanded in many other countries.

In the 1980s GM entered the small car market segment in Europe with the Opel Corsa and Vauxhall Nova, built in a new plant in Spain. GM thus became a competitor in all the major market segments, and began the growth that has made Opel/Vauxhall Europe's best-selling car line, and GM Europe its most profitable volume vehicle manufacturer.

In the early 1970s, when global economies were buffeted by two petroleum crises, GM embarked on an unprecedented program to redesign its entire North

American vehicle lineup for better fuel economy. Virtually all of GM's car lines were redesigned from body-and-frame, rear-wheel-drive, to integral-body, front-wheel-drive designs. Weight and exterior size were reduced, with vehicle interior room and comfort retained.

Then-GM Chairman Thomas A. Murphy called it "the most comprehensive, ambitious, far-reaching, and costly program of its kind in the history of our industry."

The first "downsized" cars were GM's 1977-model full-size autos — about a foot shorter and 700 pounds lighter than their predecessors. They proved an instant hit and were followed by redesigned 1978-model intermediates, 1979-model personal luxury cars, 1980-model front-wheel-drive compacts, 1981-model front-wheel-drive subcompacts, 1982-model front-wheel-drive mid-size models, and the U.S. industry's first compact truck.

As the '80s began and Roger B. Smith became chairman, GM faced the challenges of modernization. Responding to customers' continuing demands for more fuel-efficient vehicles at reasonable prices, the company launched an unprecedented $40 billion, five-year capital spending program which opened the way for dramatic technological progress throughout General Motors. New auto assembly plants were opened in Orion Township and Detroit/Hamtramck, Mich.; Wentzville, Miss., and a new truck assembly center opened in Fort Wayne, Ind. Other plants were extensively modernized.

COMING HOME

In 1981, General Motors Acceptance Corporation and Motors Insurance Corporation, the Corporation's finance and insurance operations, moved their headquarters from New York to Detroit.

Later that year, in a major realignment of GM's worldwide truck and bus operations, the truck and bus group took on responsibility for the design, engineering, manufacturing, sales and service of all General Motors trucks, buses and vans.

A major project involving GM's Buick Motor Division began in 1983. It involved a complete revamping of the car assembly operations in Flint to produce all-new front-wheel-drive automo-

biles for the 1986 model year. Referred to as "Buick City," the concept also involves supplier firms and ultimately created nearly 5,000 jobs in the integrated complex.

In 1984, GM restructured its North American Passenger Car Operations into two integrated car groups. The two groups, Chevrolet-Pontiac-GM of Canada (C-P-C) and Buick-Oldsmobile-Cadillac (B-O-C), had complete responsibility for engineering, manufacturing, assembly, and marketing of their products.

A new milestone for the U.S. automobile industry was reached in February 1984, when GM and Japan's Toyota Motor Company formed a joint venture to produce a new small car in a plant which GM had previously closed in Fremont, Calif. The joint venture company was named New United Motor Manufacturing, Incorporated (NUMMI), and its first automobile came on the market June 13, 1985, as the Chevrolet Nova.

In a transaction completed Oct. 18, 1984, General Motors acquired Electronic Data Systems of Dallas, Texas. Operating as an independent consolidated subsidiary, EDS, has grown as a world leader in the information technology industry. It serves GM through more effective management control of the company's data processing, business information and communications services, as well as serving its continually growing list of worldwide business and government clients.

Also in 1984, GM added Saturn to its vehicle divisions, thus adding a sixth automobile brand. Saturn Corporation is headquartered in Troy, Michigan, and produces its cars in a highly integrated plant at Spring Hill, Tenn., 30 miles south of Nashville. Saturn's innovative production, management, personnel and retailing policies have produced a family of subcompacts that compete with Japanese imports and a business philosophy that achieves industry-leading customer satisfaction levels.

In mid-1985, GM diversified and expanded its capabilities in advanced technology when it purchased Hughes Aircraft Company, based in Los Angeles, Calif., for $5 billion in cash and securities. This acquisition accelerated the rate of application of electronics into GM's automotive products, and provided GM with world-renowned systems engineering resources. GM combined its Delco Electronics Division and some other GM technical

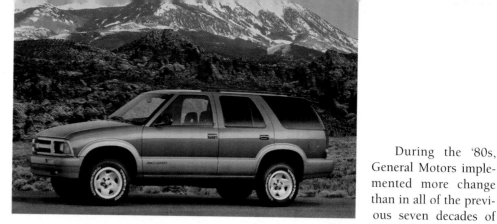

1995 Chevrolet Blazer sport/utility vehicle, one of GM's most popular and successful vehicles of the mid-decade. (General Motors)

units with Hughes to create a new subsidiary, GM Hughes Electronics Corporation (GMHE). (Renamed Hughes Electronics Corporation in March, 1995.)

SLIMMING DOWN

Recognizing it would face many challenges going into the 21st century, confronting foreign competition and a global economy, GM embarked on a tough cost-reduction course. The year 1986 began a period of plant closings, significant reduction of its worldwide workforce, phasing out of noncompetitive or obsolete component manufacturing operations, and overall reduction of operating expenses. Together with the opening of six new plants and investments to retool and modernize other facilities, these efforts were designed to make General Motors a cost-competitive producer of high quality products.

GM's commitment to quality was intensified in 1987 when GM executives and leaders of the United Automobile Workers (UAW) formed the Quality Network, a joint effort for strategic development and production of high quality, customer-valued products.

In October 1987, General Motors and the UAW signed a historic three-year labor agreement that underscored a new spirit of teamwork and partnership between management and labor. The agreement, reached without a work stoppage or strike deadline, included unprecedented job security provisions and established joint study committees at GM plants around the U.S. to review operational competitiveness and find ways to improve quality and efficiency. The job security provisions were augmented in successive contracts in 1990 and 1993.

In 1987, GM's Sunraycer, a solar powered car designed and built by several GM units, won the inaugural transcontinental World Solar Challenge race in Australia, outdistancing its nearest competitor by more than 600 miles. It's flawless performance over the six-day trek was further evidence of GM's leadership in new automotive technologies, including electric propulsion, lightweight materials and aerodynamics.

On the global front, GM purchased 50 percent of Saab Automobile AB of Sweden in 1989. GM's International Operations now work with Saab to develop, manufacture and market Saab passenger cars worldwide.

During the '80s, General Motors implemented more change than in all of the previous seven decades of the Corporation's history — with new plants, new technology, new products, a new commitment to cost efficiency, and a new commitment to its people.

In the '90s, it is apparent that all automobile manufacturers are affected by economic uncertainties, competitive pressures, intense global competition, stringent fuel economy demands, tougher emissions standards, and a pace of change more challenging than ever.

In 1991, the American automotive industry sustained financial losses unparalleled in its history. The challenges facing GM were particularly acute, primarily in the North American automotive operations, and despite strong revenues and earnings of its subsidiaries, GM recorded overall losses in 1990, 1991 and 1992.

GM accelerated fundamental changes in its ways of doing business. Plants were idled, the salaried and hourly work force was reduced through attrition and retirements, executive compensation was reduced and many non-core assets were sold.

Customer satisfaction became an overriding concern. Every division established 24-hour Roadside Assistance programs, Bumper-to-Bumper Plus warranty covered every part of every GM car and light truck for three years or 36,000 miles, and GMAC's Smart Lease program was introduced to offer customers the option to lease a GM vehicle with typically lower monthly payments.

The year 1992 was one of major management changes at General Motors. GM launched a major reorganization to streamline its business practices and turn around its North American Operations (NAO), which replaced the C-P-C and B-O-C Groups. These changes were focused to achieve GM's vision of total customer satisfaction and restore profitability. GM's new structure led to more flexible decision-making processes, more efficient use of technical and capital resources, and management accountability for production of high-value, high-quality products and services.

CENTRALIZED EXPERTISE

The GM Technical Center consolidated its staffs to become the NAO Technical Center. A centralized Vehicle Launch Center (VLC) was formed to concentrate GM's engineering and technical resources for developing new products; engineers from the car and truck divisions were joined by engineers from the Engineering Center and the Manufacturing Center to work as a team.

GM's component manufacturing divisions were consolidated into the GM Automotive Components Group Worldwide (ACGW) (renamed Delphi Automotive Systems in February, 1995) creating the largest supplier organization in the industry. Its focus became global, and its business units that did not have growth or profit potential were closed or sold.

Responding to the competitive realities of the marketplace, GM acted to restore the Corporation's operations and financial structure for long-term health. The objective was fundamental change, with minimum disruption, to eliminate redundancies, focus on value-added activity, and improve the overall responsiveness of the organization.

A variety of approaches were used to make significant reductions in both the salaried and hourly work forces, while still providing an effective safety net for displaced employees. The Central office staff was reduced from about 13,500 to about 2,300 with many of the functions transferring to operating units.

Among successful marketing innovations, GM launched the "GM Card" in 1992. It is a Mastercard credit card which enables users to build up annual credits of 5 percent or more on each item charged, and apply them toward the future purchase or lease of a new GM vehicle.

GM's difficulties in the past few years were in a sense an overdue wake-up call for the company. Its long period of success had made it easy to ignore the significance of change and the signs of potential future problems, as the corporation's legendary leader, Alfred Sloan, had warned could happen when he published his memoirs in 1963. The lesson is that, for unrivaled leaders, success itself can breed complacency, myopia and ultimately, decline. Other leaders in several industries and businesses have had the same harsh wake-up call in recent years.

The most urgent challenge was to reverse the financial losses from the North American Operations. In 1993, intensified efforts yielded results in the areas of customer focus, product quality, global sourcing and advance purchasing, lean manufacturing, commonization of processes, systems, and parts, and integration of global resources.

And in 1993, GM and Toyota signed an unprecedented agreement for GM to supply North American-built cars to Toyota for sale in Japan. GM will build right-hand drive Chevrolet Cavaliers in the U.S., which Toyota will purchase and sell in Japan.

The year 1993 was a watershed one in GM's drive to return to profitability and reassert industry leadership. Reflecting a major improvement in North American Operations (NAO) as well as strong earnings from International Operations, GMHE, EDS and GMAC, the corporation earned a total of $2.5 billion, a dramatic and gratifying turnaround after three straight years of staggering losses.

RECORD EARNINGS

In 1994, GM recorded all-time record net income of $4.9 billion, and all of GM's business sectors reported strong sales and earnings. Its success was spread across all its business sectors and geographical operations; North American Operations were profitable, and GM Europe was that region's most profitable volume auto manufacturer.

EDS and Hughes Electronics also reported record earnings in 1994 and strengthened their positions in fast-moving technologies of information management and telecommunications. Hughes' launch of DIRECTV was the most successful new product introduction in consumer electronics history.

Automotive Component Group Worldwide (ACGW) (now Delphi Automotive Systems), became a separate business sector of the Corporation in 1994. With sales of over $26 billion, and 190 operations and 17 technical centers in 31 countries, it is the world's largest automotive systems manufacturer. Growing rapidly, in the '90s, it has established manufacturing operations, customer service offices and joint ventures in China, Japan, Europe and Australia. It is now a supplier to virtually all the world's automobile manufacturers, as well as a strategic partner to GM's vehicle operations.

Looking to the future, GM is in transition from a base of multinational and regional operations to consolidated global strategies. Planning is under way to coordinate many of the North American and International vehicle platforms, the common structural systems which are the basis of its cars and trucks. With common engineering and manufacturing systems and common components, GM will be able to offer a greater variety of vehicles tailored to needs and tastes of customers in the various worldwide markets and build them with lower costs.

As John F. Smith, Jr., GM chief executive officer and president stated, "GM is changing its ways and will continue changing."

When it comes to cars, we're like a kid.

It's pure joy. At Budd, the very idea of getting an automotive program geared up and rolling toward
introduction lights a spirit and enthusiasm you won't find in any other company. That's because our greatest thrill
lies in knowing that we can do it. Concepting. Design. Engineering. Prototyping. Testing. Pre-production validation
and manufacturing. Budd has the capability, the facilities and the bright eager minds to carry a program through every
step of the way. It doesn't matter whether the need is steel, frames, sheet molded composite, castings or cold-weather
starting products. There's enough state-of-the-art technology and fresh new thinking at Budd to make anyone
who loves cars feel just like a kid again. Phone Budd at (810) 643-3520 and join the excitement.

THE **Budd** COMPANY

3M salutes innovation

the engine of progress for over 100 years.

1965 Ford Mustang 2+2 "fastback" symbolized the muscle cars of the 1960s and 70s. (National Automotive History Collection, Detroit Public Library)

Ford:
On The Fly

*Moving assembly line is
one of the most widely recalled
of many innovations*

Ford Motor Company entered the business world without fanfare on June 16, 1903, when the late Henry Ford and 11 associates filed incorporation papers at Michigan's State Capitol in Lansing. With an abundance of faith, but only $28,000 in cash, the pioneering industrialists gave birth to what was to become one of the world's largest corporations.

Few companies are as closely identified with the history and development of America throughout the 20th century as Ford. And perhaps no other American firm is as well known across the globe.

At the time of its incorporation, Ford was a tiny operation in a converted Detroit wagon factory staffed with 10 people. Today, the company is the world's fourth-largest industrial corporation and the second-largest producer of cars and trucks, with active manufacturing, assembly or sales operations in 31 countries on six continents.

Some 338,000 men and women now come to work each day in Ford factories, laboratories and offices around the world. Ford products are sold in more than 200 nations and territories by a global network of some 10,500 dealers. And the company's annual sales exceed the gross national products of many industrialized nations. For some 30 years, Ford has sold more vehicles overseas than any other U.S.-based manufacturer.

Like most great enterprises, Ford's beginnings were modest. The company had anxious moments in its infancy, balancing the scale of bankruptcy precariously.

EVEN IF YOU DON'T GO TO THE INDY 500, YOU CAN STILL GET THE SOUVENIRS.

NEW Firehawk SS10   **NEW** Firehawk Touring LH   Firehawk GTA   Firehawk SVX   Firehawk SZ

Because 1995 marks the year we bring them to you as the Legend of Firestone returns to the prestigious Indianapolis 500. From the same kind of extensive testing that went into making that return possible comes the Firestone Firehawk, a full line of street performance radials speed rated from S to Z, engineered for all-around handling and performance. So if you demand quick acceleration, grip, and stability at high speeds, stop by your local Firestone retailer and check out Firehawk today. After all, if Firehawks can perform at 240 miles per hour at the Indy 500, imagine how well they'll do for you on the highway.

FIREHAWK
AMERICA'S PERFORMANCE TIRE

Firestone
America's Tire Since 1900

The Indianapolis 500®, Indy 500® and IndyCar™ are trademarks of the Indianapolis Motor Speedway. IndyCar under exclusive license to Championship Auto Racing Teams, Inc.

But one month after incorporation a ray of hope shone when the first car was sold to a Chicago dentist named Pfennig. A worried group of stockholders, skeptically eyeing a bank balance that had dwindled to $223.65, breathed easier, and a young Ford Motor Company had taken its first step.

During the next five years, young Henry Ford, as chief engineer and later as president, directed a development and production program which started in a converted wagon factory on Mack Avenue in Detroit and later moved to a larger building at Piquette and Beaubien Streets. In the first 15 months 1,700 Model A cars chugged out of the old wagon factory.

Between 1903 and 1908, Henry Ford and his engineers used the first 19 letters of the alphabet to designate their creations, although some of the cars were experimental and never reached the public.

The most successful of the early production cars was the Model N — a small, light, four-cylinder machine which went on the market at $500. A $2,500 six-cylinder luxury car, the Model K, sold poorly.

The Model K's failure, along with Mr. Ford's insistence that the company's future lay in the production of inexpensive cars for a mass market, caused increasing friction between Mr. Ford and Alexander Malcomson, a Detroit coal dealer who had been instrumental in raising the original $28,000. As a result, Mr. Malcomson left the company and Mr. Ford acquired enough of his stock to increase his holdings to 58-1/2 percent. He became president in 1906, replacing John S. Gray, a Detroit banker.

FINANCIER FACTIONS

Squabbles among the stockholders did not threaten the young company's future as seriously as did a man named George Selden. Mr. Selden had a patent on "road locomotives" powered by internal combustion engines. To protect his patent, he formed a powerful syndicate to license selected manufacturers and collect royalties for every "horseless carriage" built or sold in America, attempting to monopolize the industry.

Hardly had the doors been opened at the Mack Avenue factory when Selden's syndicate filed suit against the Ford company which bravely had gone into business without a Selden license.

Faced with the choice of closing the doors or fighting a battery of attorneys who already had whipped big-

ger companies into line, Henry Ford and his partners decided to fight.

Eight years later, in 1911, after incredibly complicated court proceedings, Ford Motor Company won the battle which freed it and the entire auto industry from Selden's strait jacket.

Despite harassments from Selden's syndicate, the little company kept improving its machines, making its way through the alphabet until it reached the Model T in 1908. A considerable improvement over all previous models, this car was an immediate success.

Nineteen years and more than 15 million Model T's later (1927), Ford Motor Company was a giant industrial complex that spanned the globe. Its cars had started an urban revolution. Its $5 day and the philosophy behind it had started a social revolution. Its moving assembly line had started an industrial revolution.

Henry Ford shown in his first experimental car, the Quadricycle, run on Detroit streets June 4, 1896. The Model T didn't arrive for another dozen years. (National Automotive History Collection, Detroit Public Library)

Tonight, a man works the midnight shift, a doctor performs emergency surgery, and a mother reads while waiting up for her son.

This could be any city, it happens to be one of ours.

Detroit Edison

Some of the world's leading hospitals, universities, manufacturers and research facilities rely on us for one thing – electricity. At Detroit Edison, we are committed to delivering the highest quality and most reliable electric service for all the important work done in our community.

During those years of hectic expansion, Ford Motor Company:

- began producing trucks and tractors (1917);
- became wholly owned by Henry Ford and his son, Edsel, who succeeded his father as president in 1919 after a conflict with stockholders over the millions to be spent to build the giant Rouge manufacturing complex in Dearborn, Mich.;
- bought the Lincoln Motor Company (1922); and
- built the first of 196 Ford Tri-Motor airplanes used by America's first commercial airlines (1925).

By 1927, time had run out on the Model T. Improved, but basically unchanged for so many years, it was losing ground to more stylish, powerful machines offered by Ford's competitors. On May 31, Ford plants across the country closed down for six months while the company retooled for the Model A.

It was a vastly improved car in all respects. More than 4,500,000 Model A's, in several body styles and a wide variety of colors, rolled onto the nation's highways between late 1927 and 1931.

But even the Model A eventually reached the end of the line when consumers demanded more luxury and power. Ford Motor Company delivered both with its next entry — the Ford V-8 — which the public saw

Ford's Highland Park plant about 1915, the home of the Model T, the moving assembly line and the $5 day. Note streetcar terminal in foreground on west side of Woodward Avenue. In 1995, the administration building — abandoned and much-neglected — and factory buildings at the right rear used largely for warehousing still stood. (Albert Kahn Associates) **BELOW:** *On July 3, 1945, Ford was the first automaker to resume civilian production. Here young Henry Ford II, released from Navy duty in 1943 to help manage the company, presents keys to that first postwar car to President Harry S. Truman. The presentation was merely symbolic. When Truman relinquished the presidency in 1953, he still owned the 1941 Chrysler Royal two-door sedan purchased new when he was in his second term as U.S. Senator from Missouri. (Mike Davis collection)*

for the first time on April 1, 1932. Ford was the first company in history to cast successfully a V-8 engine block in one piece. Experts told Mr. Ford it couldn't be done. It was many years before Ford's competitors learned how to mass-produce a reliable V-8. In the meantime, the car and its powerful engine became the darling of performance-minded Americans.

In 1938, six years after the V-8 was introduced, production started on the Mercury which became Ford's entry in the growing medium-priced field.

In 1942, civilian car production came to a halt as the company threw all its resources into the U.S. war effort. Initiated by Edsel Ford, this giant wartime program produced in less than three years 8,600 four-engined B-24 "Liberator" bombers, 57,000 aircraft engines, over 250,000 jeeps, tanks, tank destroyers, and other pieces of war machinery.

REVERSE SUCCESSION

Edsel Ford died in 1943 just as his program was reaching its maximum efficiency. A saddened Henry Ford resumed the presidency until the war's end when

he resigned for the second time. His oldest grandson, Henry Ford II, became president on Sept. 21,1945.

Even as Henry Ford II drove the industry's first post-war car off the assembly line, he was making plans to reorganize and decentralize the company. Losing money at the rate of several million dollars a month, Ford Motor Company was in critically poor condition to resume its prewar position as a major factor in a fiercely competitive auto industry. In much the same manner as his grandfather had faced problems at the company's beginning, Henry Ford II tackled the job of building an automobile company all over again. His postwar reorganization and expansion plan rapidly restored the company's health.

Having relinquished his company's operation to his grandson, Henry Ford lived quietly with his wife, Clara, until he died on April 7, 1947, at age 83.

Another important figure in the company's history in the post-war years was Ernest R. Breech who served as chairman from Jan. 25, 1955, until he was succeeded by Henry Ford II on July 13,1960.

Paralleling Ford's domestic growth has been a foreign expansion program which began in 1904, just one year after the company was formed. On Aug. 17 of that year, a modest plant opened in the small town of Walkerville, Ont., with the imposing name of Ford Motor Company of Canada, Ltd.

Ford's standard pickup truck, for many years the largest-selling vehicle model in North America, if not the world. (Mike Davis collection) TOP: 1995 Ford Taurus sedan, the nation's top selling passenger car for several years running, and a styling leader for nearly a decade. (Ford Motor Company)

From this small beginning has grown an overseas organization of manufacturing plants, assembly plants, parts depots and dealers, with Ford represented in some 200 countries and territories around the world. About 60,000 companies worldwide supply Ford with goods and services.

Another significant event in the company's history occurred in January 1956, when Ford common stock was sold to the public for the first time. The company now has some 287,000 stockholders.

On Oct. 1, 1979, Henry Ford II retired from the position of chief executive officer, handing over that responsibility to Philip Caldwell. Mr. Caldwell also succeeded Mr. Ford as chairman of the Board on March 13, 1980. Mr. Ford retired as an officer and employee of the company on Oct. 1, 1982, but continued to serve on the company's Board of Directors and as chairman of the Finance Committee until his death on Sept. 29,1987.

When Philip Caldwell retired as chairman of the board and chief executive officer, Donald E. Petersen succeeded him in that position Feb. 1,1985. When Mr. Petersen retired March 1, 1990, he in turn was succeeded by Harold A. Poling.

Ford Motor Company is the second-largest car and truck producer in the world and is ranked third on the Fortune 500 list of the largest U.S. industrial corporations, based on sales. In 1991, Ford's worldwide sales and revenues totaled $88.3 billion.

Although Ford is better known as a manufacturer of cars, trucks and tractors, it now produces a wide range of other products including industrial engines, construction machinery, glass and plastics. And Ford is established in a diversity of other businesses including financial services, insurance, automotive replacement parts, electronics and land development.

HADEN, INC. PROCESS ENGINEERING SOLUTIONS

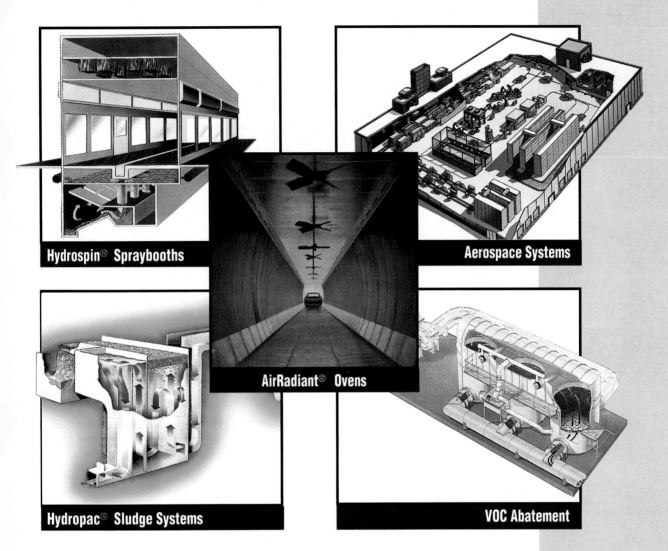

Hydrospin® Spraybooths

AirRadiant® Ovens

Aerospace Systems

Hydropac® Sludge Systems

VOC Abatement

Haden, inc. is a wholly-owned subsidiary of Haden MacLellan Holdings plc which has serviced industry since 1890. Throughout our relationship with the automotive industry we have taken great pride in the technical advances we have made to help improve our customers' product quality and productivity. Haden has been the originator of key advances such as the glass walled center downdraft Hydrospin® spraybooth, Hydropac® flotation cell sludge system, and DryPure™ paint sludge dryer. Haden continues to be the worldwide technology leader with high efficiency Hydrospin III spraybooths, Air Radiant® ovens, and automotive quality powder paint systems.

The Chrysler Story

Last, smallest of the Big Three — but not the 'least'

TOP: *Walter P. Chrysler checks over one of his brand new, snappy-looking 1924 Chrysler roadsters in the summer of that year. Chrysler, a former General Motors and Willys executive, had been brought in by bankers and investors to save the ailing Maxwell-Chalmers car company. (Chrysler Historical Collection)*

When Walter P. Chrysler founded Chrysler Corporation in 1925, the Chrysler Technology Center was probably beyond even his wildest dreams. He could hardly have imagined a $1 billion complex where teams of designers, engineers and manufacturers work together in 3.5 million square feet of the most sophisticated laboratories and testing facilities in the world.

The Technology Center, in Auburn Hills, Michigan, represents the next era of Chrysler's history. Chrysler is poised as a lean, bold, quality-focused company where teams of employees develop and build cars and trucks totally focused on customer needs. Teamwork will lead the company into the 21st century, but it was the vision and ambition of one man more than 70 years ago that began the Chrysler Corporation story.

Chrysler Corporation, founded on June 6, 1925, has a history as rich and colorful as that of its founder, Walter P. Chrysler — a man so intrigued with automotive technology that he bought, disassembled and reassembled an automobile before he learned to drive.

Walter Chrysler was born in Wamego, Kansas, on April 20, 1875. He began an ambitious career in the railroad industry as a machinist's apprentice when he was 17. Chrysler earned his master mechanic's papers in 1899 and nine years later, at the age of 33, became the youngest man ever to hold the post of superinten-

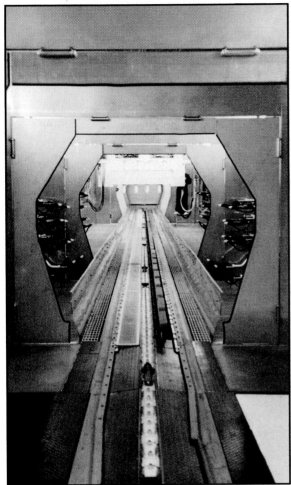

dent of motive power for the Chicago Great Western Railway.

When he joined the American Locomotive Company as manager of its Pittsburgh Works in 1910, his curiosity about the infant automobile industry led him to buy his first automobile, a Locomobile Phaeton.

Two years later, Chrysler turned his fascination for automobiles into a career by joining the Buick Motor Car company in Flint, Michigan, as works manager. Buick became General

Motors' first automotive division when GM incorporated in 1916. A year later, Walter Chrysler was named the division's president and general manager.

RETIRED, THEN REHIRED

In 1919, in addition to his responsibilities at Buick, Chrysler became General Motors' first vice president in charge of manufacturing. Financially independent at age 45, he retired from General Motors in 1920. But, within a year he was wooed out of retirement to become executive vice president of Willys-Overland which was in financial trouble. He was later hired to aid the ailing Maxwell Motor Car Company, Inc.

He was named chairman of Maxwell Motor Car's Reorganization and Management Committee and revitalized the company with the development of the Chrysler Six — America's first medium-priced, high-styled automobile. By January 1924, Maxwell Motor Car had set an industry sales record when sales of the Chrysler Six reached 32,000 units.

Chrysler Corporation was incorporated in Delaware on June 6, 1925, as a successor to the Maxwell Motor Car. Walter Chrysler became chairman of the board and president. In 1925, the company introduced the Chrysler Four, Series 58, which had a top speed of 58 mph. More than a million people visited showrooms in the first four days the corporation displayed the new automotive wonder. By the fall of 1925, more than 3,800 dealers were selling Chrysler cars and,

Introduction of the Dodge Brothers' new car in Boston in 1914. John and Horace Dodge became millionaires on Ford stock received in exchange for supply of major parts for early Ford cars, and launched their own make very successfully. Both died in 1920, and Chrysler acquired the company from Wall Street bankers in 1928. (Chrysler Historical Collection) **BELOW:** *In a three-month period in the summer of 1928, Chrysler Corporation launched the low-priced Plymouth car, acquired Dodge and introduced the middle-priced DeSoto. In the 1930s and '40s, specially built DeSoto cabs became almost the standard in large American cities. These 1936 De-Sotos were ready for a Yellow Cab driveaway from the factory, probably at Wyoming and Ford Road. (Chrysler Historical Collection)*

ing the first car to drive into and out of the Grand Canyon.

By 1929, Chrysler had grown to become one of the Big Three leading automotive manufacturers. The company endured the Great Depression of the '30s through cost-cutting measures but never cut back on research and development. That decision paid off in 1933 when Chrysler became the only automotive company whose sales surpassed the record 1929 sales year.

Chrysler built its reputation around a trio of engineering "greats; shown here are Fred Zeder, second from left, and Carl Breer, at far right, with the streamlined Airflow introduced for 1934. The third member of the trio, Owen Skelton, is not pictured. (Chrysler Historical Collection) BELOW: 1953 Dodge Coronet, a conservative technical achievement for the time — smaller on the outside, bigger on the inside — which didn't meet public acceptance. Two years later, Chrysler joined industry trends toward bigger, flashier and more powerful. (Chrysler Historical Collection)

Chrysler's commitment to R&D in the 1930s resulted in several Chrysler engineering innovations including the patented "Floating Power," an engine mounting system which successfully isolated engine vibration away from the passengers and resulted in a smoother ride.

In 1934, Chrysler introduced the Chrysler Airflow and De Soto Airflow vehicles, radically styled for their time with the industry's first one-piece curved glass windshield. Engineered for a smooth ride, the Airflow was equipped with Chrysler's first Automatic Overdrive Transmission.

Chrysler continued to expand its operations by establishing Chrysler de Mexico in 1938 as an importer and distributor of Chrysler products. Subsequently, the company became an assembler and later a manufacturer of power trains. Chrysler de Mexico was controlled by Mexican capital until 1971 when Chrysler acquired majority ownership. Today, Chrysler de Mexico operates nine facilities in Mexico, including two vehicle assembly plants.

Walter Chrysler led the corporation through the early years of innovation and North American expansion until he resigned the presidency on July 22, 1935. He remained chairman of the board until his death on Aug. 18, 1940. He is remembered as an inspired entrepreneur and an early patriarch of America's automotive industry.

by 1926, the corporation had risen from 57th to fifth place in industry sales.

Chrysler had expanded its operations to Canada as well. The Maxwell-Chrysler Company of Canada began in 1924 in Windsor, Ontario. The company incorporated as Chrysler Corporation of Canada, Ltd. in 1925. Windsor, an industrial city across the river from Detroit, quickly became Canada's largest automotive manufacturing center.

In 1928, the De Soto and the first Plymouth — priced from $670 to $725 to appeal to consumers with average incomes — went into production in the United States. That same year, Chrysler bought Dodge Brothers, Inc., automotive builders with a solid reputation for making vehicles with rugged dependability. The Dodge Brothers became famous for build-

TANKS AND TRUCKS

In the 1940s, American industry turned from consumer to defense production. Chrysler participated in many defense projects, most notably production of the 32-ton Sherman M4 tank. Chrysler produced 18,000 Sherman tanks, the main combat vehicle of the U.S. and its Allies in World War II. Approximately 500,000 Dodge trucks were also produced for the war. By 1945, Chrysler had supplied more than $3.4 billion in military equipment to the Allied forces.

After the war, Chrysler responded to the need for increased car and truck production by building or buying 11 plants between 1947 and 1950. In the 1940s, Chrysler introduced the Town & Country sedans and convertibles and began limited production of its first hardtop convertible in 1946. In 1951, Chrysler introduced the first production power steering vehicle. The first key-operated ignition and safety-cushioned dash were other Chrysler innovations of the early '50s, as well as the Torsion-Aire suspension system in 1957.

K. T. Keller became chairman of the board on Nov. 3, 1950, as Chrysler entered the space age by signing a contract with the U.S. Army to build the Jupiter space exploration missiles. In 1952, one of America's first successful space flights carried two monkeys to an altitude of 350 miles at the top of a Jupiter missile.

When hostilities erupted in Korea in 1950, Chrysler again supplied military products, including tanks, military trucks and air raid sirens. By the end of the Korean police action, Chrysler had participated in 31 government projects worth more than $1.1 billion.

Chrysler's engineering department developed several landmark innovations during the 1950s including four-wheel, self-energizing hydraulic disc brakes and the famous "Hemi" V-8 engine.

Chrysler engineers created the hemispheric combustion chamber V-8 engine, the Hemi, in 1951. The Hemi engine was a relatively expensive option for retail customers. Its costs often equaled a third of the purchase price of the car. Chrysler stopped producing the Hemi in the mid-50s due to the costs. However, in 1964, Chrysler introduced a high-powered sport version of the engine — the legendary 426 Hemi — with more than 400 horsepower under the hood. Richard Petty made the 426 famous in his NASCAR racing. Only about 11,000 of the specialty engines were built between 1964 and 1971. Car enthusiasts still consider the Hemi to be the ultimate internal combustion engine.

Other Chrysler developments in the 1950s have become standard features on today's vehicles. The elec-

Lee A. Iacocca was educated as an engineer at Lehigh and Princeton and joined Ford as a manufacturing engineering trainee at the Rouge, later switching to sales and rising to president. After a falling out with Henry Ford II, he became chairman of Chrysler and is credited with saving the company in the early 1980s. Here he is shown pitching a 1981 Plymouth Reliant "K-car" at a press conference in 1980. (Chrysler Historical Collection) PAGE 59: 1996 Plymouth Grand Voyager, second generation of the company's pace-setting front-wheel drive, car-like minivans. Chrysler completely revamped its car and truck lines in the early to mid-1990s. (Chrysler Corporation)

WHEN YOU THINK OF ROCKWELL, YOU'VE THOUGHT OF EVERYTHING.

So just how many products and services do you think Rockwell Automotive offers to OEM's? Here's a hint: If it opens, closes, pushes, pulls or turns, we have designed, engineered and manufactured a component or system to play a part in it. In fact, you may know us for one thing and be totally surprised that we do so many other things as well. For instance, did you know that Rockwell Automotive is a manufacturer of not only suspension systems, wheels, sunroofs and automotive electronics but also of door and access control systems, seat adjusters and all kinds of heavy vehicle components? So as your business goes global, think local—with Rockwell Automotive.

 **Rockwell** *Automotive*

When you think of Rockwell, you've thought of everything.

tric window lift system (1950), power steering and Oriflow shock absorbers (1951) made driving safer and easier. Drivers of Chrysler products built in 1955 were the first to enjoy the pleasures of Chrysler's all-transistor car radios. And, two years later, curved side windows expanded the possibilities for automotive design.

In 1960, after 33 years, production of the De Soto car line came to an end. The same year, Chrysler introduced the 45 rpm automotive record changer with 14 record capacity. Three years later, Chrysler offered the industry's first five-year or 50,000 mile warranty on drive train components.

Chrysler experienced a number of transitions in the 1960s. On April 29, 1960, L. L. Colbert became chairman of the board, followed by George H. Love the next year on Sept. 21, 1961 and Lynn A. Townsend on Jan. 1, 1967.

In 1963, under Love's leadership, Chrysler introduced the Turbine. The vehicle was powered by a turbine engine, similar to today's jet engines.

Only 50 of the vehicles were manufactured for consumer research and testing. The Turbine never went into full production, but the research results were incorporated into defense vehicle technology.

CLEARING THE AIR

Chrysler innovation continued in 1966, when engineers modified engines to create the Clean Air Package, an exhaust emission control system. The front seat shoulder harness and the separate, self-contained rear window heater/defroster system were among Chrysler safety innovations in 1966.

As it became increasingly difficult to produce small cars in the United States profitably, Chrysler began importing and distributing small passenger cars and trucks built by its Japanese partner, Mitsubishi Motors Corporation.

Chrysler set sales records in 1972 and 1973, but gasoline shortages, political uncertainty, high interest rates, severe inflation and weakening consumer confidence drove Chrysler into a financial crisis in the mid-'70s. American consumer demand soared for smaller, more fuel-efficient cars. Japanese manufacturers were the first to respond, making great inroads into the U.S. market. The combined domestic market share of the

total U.S. car market fell while the market share for imports rose to 23.4 percent.

In the midst of the financial crisis, John J. Riccardo became chairman on Oct. 1, 1975. Responding to growing economic trouble, Riccardo hired Lee A. Iacocca as Chrysler president on Nov. 2, 1978. Ten months later, Riccardo resigned and Iacocca was elected chairman on Sept. 20, 1979.

Iacocca applied his experience of 32 years with Ford Motor Company to meet the challenge of rejuvenating Chrysler's sagging operations. Chrysler reduced costs, restructured its management and recruited new executives to deal with its serious financial problems. Despite these measures, external factors continued to limit Chrysler's ability to finance its programs fully. Chrysler was forced to seek assistance from the federal government in the form of loan guarantees.

In late Dec. 1979, the U.S. Congress passed the Chrysler Corporation Loan Guarantee Act, which President Jimmy Carter signed into law on Jan. 7, 1980. The act provided Chrysler $1.5 billion in federal loan guarantees.

Concessions from UAW-represented workers, white-collar employees, suppliers, creditors and lenders kept Chrysler operating despite record losses of $1.7 billion in 1980. Chrysler cut inventories by $1 billion, reduced white-collar staff by 50 percent and cut its break-even point by 50 percent in its drastic efforts to manage finances.

Through the travail, Chrysler doubled its fleet average miles-per-gallon and in 1978, introduced the first domestically produced front-wheel drive small cars — the Dodge Omni and Plymouth Horizon. Chrysler was also the first American company to convert its fleet to front-wheel drive. Chrysler was on its way to recovery.

"If you can find a better car ... buy it." This challenge became Chrysler's battle cry in its recovery fight. Iacocca began appearing in Chrysler's advertising in July 1980 and became one of the most recognizable businessmen in the world.

In 1981, Chrysler reported record losses, but the company saw light at the end of its financial tunnel — from the headlamps of its new K-cars. Developed on a limited budget, the Dodge Aries and the Plymouth

Reliant, code-named the "K-cars," enjoyed sales success, which Chrysler rode to profitability in 1982. The momentum continued, and for the first time since 1973, the company was profitable for four consecutive quarters. In August 1983, Chrysler paid off the federal loan guarantees seven years early, at a profit of $350 million to the U.S. government.

MINIVAN MADNESS

In November 1983, production of Chrysler's minivans, the Dodge Caravan and Plymouth Voyager, began. The minivans created a new market segment and changed the way American families traveled. Minivans became Chrysler's best-selling vehicle and the company was well on its way back to economic health. More than a decade later, despite an onslaught of domestic and international minivan competition, Chrysler continues to dominate the U.S. minivan market and has captured more than 20% of Europe's minivan market.

Chrysler expanded into electronics and aerospace activities and enlarged its international operations in the 1980s. In 1984, the company reported its best earnings year ever and reorganized itself into a holding company made up of four operating divisions — Chrysler Motors, Chrysler Financial, Gulfstream Aerospace and Chrysler Technologies.

Chrysler continued its overseas expansion as it acquired 15.6 percent equity in Italian luxury car manufacturer, Officine Alfieri Maserati SpA, in 1984. The relationship between Maserati and Chrysler ended in August 1988 and the equity was later sold.

Chrysler and Mitsubishi Motors Corporation formed Diamond-Star Motors Corporation in 1985, as a joint venture company to manufacture small cars in the United States. Production of the Plymouth Laser started three years later in Normal, Illinois.

After a nine-year absence, Chrysler re-entered the European market in 1987 by exporting the Chrysler LeBaron convertible, the Plymouth Voyager and the Dodge Daytona. All the vehicles were sold in Europe under the Chrysler badge. Later that year, Chrysler Motors purchased Nuova Automobili F. Lamborghini SpA, the maker of the famous Countach. Today, the Italian specialty automotive manufacturer produces the Diablo — the world's fastest production automobile — as well as other sports cars and high performance marine racing engines. In 1993, Chrysler sold Nuova Automobili F. Lamborghini SpA and its subsidiaries to MegaTech Ltd.

Chrysler also made its biggest acquisition in 1987, purchasing American Motors Corporation, the fourth largest U.S. automotive company. The $800 million acquisition included the world-famous Jeep, three automotive assembly plants, 1,600 dealerships, and a joint venture, Beijing Jeep Corporation of Beijing, China.

As a result of the acquisition, Chrysler launched the Eagle brand in 1987, the first new Chrysler brand since 1927 when the Dodge brand was launched.

Chrysler continued its commitment to value and safety throughout the 80s. In 1987, Chrysler introduced an unprecedented 7-year/70,000 power train warranty and a 7-year/100,000-mile outer body rust protection warranty. The continuous-flow fully electronically-controlled fuel injection engine was a key product development in the early '80s. Chrysler became the first company to offer air bag restraint systems as standard equipment, in 1988.

In the summer of 1989, Chrysler began a $1 billion cost cutting and restructuring program to focus its resources on its core automotive business.

The restructuring led Chrysler to begin a new approach to car and truck production called "platform teams" where representatives from various departments — such as design, engineering, purchasing, manufacturing and marketing — work together on a single vehicle line through its entire life cycle. Each team functions like a small company with total operating responsibility. The team approach cuts development time, because everyone works together from the start.

The Dodge Viper, a V-10 roadster, was the first vehicle developed by a platform team. Following the successful development of the Viper platform were the 1992 Jeep Grand Cherokee and a new line of family sedans with innovative cab forward design: the 1993 Chrysler Concorde, Dodge Intrepid and Eagle Vision.

Iacocca dedicated the Chrysler Technology Center in 1991, a 3.5 million square-foot megastructure on a site where the corporation's new world headquarters is scheduled to be completed in 1995. The facility supports cross-functional work among product design, engineering, manufacturing and other departments in vehicle development.

URBAN INVESTMENT

In 1992, Chrysler dedicated the Jefferson North Assembly Plant in downtown Detroit, a $1.6 billion investment, including product development costs, to manufacture Jeep Grand Cherokee sport-utility vehicles. Chrysler built the plant in Detroit's inner city to offer continued employment to its dedicated workforce and support the city, rather than following an industry trend toward building new plants in rural sites.

The year 1992 brought a changing of the guard at Chrysler. On March 16, the board of directors named Robert J. Eaton as vice chairman and chief operating officer. Lee Iacocca stepped down as chairman and CEO on Dec. 31, 1993, and the board elected Robert Eaton to fill the posts effective Jan. 1, 1993.

In 1994, Chrysler set a company record for U.S. retail sales and earned more money — $3.7 billion — than in any other year in the company's history, including 1993, the previous record. That year also marked the introduction of the new sedans — the Chrysler Cirrus and Dodge Stratus — and the new coupes — the Chrysler Sebring and Dodge Avenger. The Cirrus was named Motor Trend's "Car of the Year." In 1995 Chrysler introduced the next generation of minivans — the 1996 minivans with the industry's first driver's side sliding door.

Today, Chrysler's aim is to become the premier car and truck company in the world by the year 2000. Chrysler continues its aggressive five-year product development strategy to produce a product line for the second half of the decade that's just as exciting as the first.

Automotive Milestones

How internal combustion
vehicles avoided the scrap heap of history

by Michael J. Kollins

Since its inception in 1893 when the first American gasoline automobile was built, almost every year of progressive development is marked by an outstanding "first" in the automobile industry. During the first five years (1893-1898) progress was slow, but after Winton's first delivery of a car to a purchaser in 1898, the public became interested. Development then became rapid. Each year thereafter was marked by some historic engineering advance or innovation of worth, except for the war years of 1917-1918 and 1942-1945.

The accompanying listing gives the dates of the most important occurrences in the development of the automobile in North America, and the introduction of the more influential features in automobile design. The dates of the origin of different practices in design in some cases may be somewhat indefinite. Any new device or practice is first conceived by the engineer, designer or inventor, and finally may be incorporated in regular production vehicles. These stages naturally are reached at different times. In preparation of this list an endeavor has been made, unless otherwise stated, to show the year when any given feature was first incorporated in cars that were sold to the public. After the mid-1930s, dates generally are for model year application.

Of course no listing of this kind can ever be 100 percent complete or 100 percent accurate; however, it is believed that most of the major historical events of lasting interest and most of the important features of present-day car construction are included.

This material was compiled by certain members of the SAE Automotive History Committee, under the leadership of Michael Kollins, from *Horseless Age, Motor Age, Motor World, Automotive Industries, Automobiles of America* and other publications. While every effort was made to make this list as full and accurate as possible, it is not infallible but rather subject to corrections, revisions, and additions.

YEAR	INVENTOR /COMPANY	DEVELOPMENT/INNOVATION
1769	Nicholas Joseph Cugnot (France)	Steam-powered traction vehicle for military transport
1804-05	Oliver Evans (United States)	Steam-powered massive dredge moved through the streets of Philadelphia and down the banks of the Schuylkill River. The driving wheels gave way to paddle wheels to propel the dredge upstream. (Perhaps the first amphibious vehicle.)
1821	Richard Trevithick (England)	Steam-powered vehicle featuring transmission with gears
1828-30	Goldsworthy, Gurney & Walter Hancock (England)	Steam-powered coaches in regular service between cities
1835	Alfred Drake (Philadelphia)	Began 20 years' experiments with internal combustion engine
1844-45	Stuart Perry (New York City)	Constructed turpentine-fueled, 2-stroke-cycle engine
1860-62	Jean Joseph Etienne Lenoir with Alphonse Beau de Rochas (France)	Four-stroke-cycle hydrocarbon benzine engine installed in road vehicle,1862. Traveled several times between Rue de la Roquette in Paris and Vincennes. Lenoir's engine was equipped with the prototype of the modern sparkplug and distributor. The German publication "Zur Frage Freien Concurrenz in Gasmotorenbaue" acknowledged the success of the vehicle.
1865	Siegfried Marcus (Austria)	Hydrocarbon-gas powered engine and vehicle; did not patent
1873	M. A. Bollee (France)	L'Obeissant steam-powered coach for 12 passengers
1883	DeLamarre-DeBoutteville (France)	With the assistance of M. Malandin, successfully constructed and operated vehicle with 4-stroke-cycle engine equipped with what was believed to be the first carburetor. Prior to this all hydrocarbon engines were fuel injected or wick capillary action.
1885	Gottlieb Daimler (German) with August Otto and Eugene Langen	Founded famous Deutz factory to construct Otto (4-stroke) cycle engine. Daimler built his first car in Cannstatt and patented the first carburetor. German records and patents also seem to confirm that the first car to leave the Karl Benz factory in Mannheim was in December 1885
1893	Charles E. & Frank Duryea (U.S.)	"Buggyaut," first appearance of gasoline-powered horseless carriage in U.S., Springfield, MA
1894	Elwood Haynes (U.S.)	Tested first car in collaboration with Elmer & Edger Apperson
1895	George B. Selden	Patent issued (November 5) for a "hydrocarbon gas engine for road or horseless carriage use" — later (1911) successfully challenged on basis it was Brayton-cycle, not the Otto-cycle engine used by all motorcars by then.
	Duryea	Won first motorcar road race, Chicago to Waukegan (Nov.28)
1896	Charles Brady King (U.S.)	Tested his car on streets of Detroit. Featured 4-cyl. en-bloc engine (March 6)
	Henry Ford (U.S.)	Tested 2-cyl., 4-hp "quadricycle" in Detroit (June 4)
1897	R. E. Olds (U.S.)	Organized Olds Motor Vehicle Company in Lansing; tested first car on August 21
1898	Alex Winton (U.S.)	Made first commercial sale/delivery of vehicle (March 24)
	Elwood Haynes	First U.S. manufacturer to use aluminum alloy
	John Wilkinson	Designed and built first 4-cyl. air-cooled engine; later used in Franklin on 1902 models
1899	R. E. Olds	Organized Olds Motor Works (May 8) , moved from Lansing to Detroit
	Packard	First Packard built (November 6)

1900	Buffalo Engine	Overhead camshaft engine, Buffalo four cylinder
	Madison Square Garden	First auto show - New York
	Madison Square Garden	First auto race - Old Madison Square Garden
	R. E. Dietz	First kerosene lamps for automobiles, 20 c.p.
1901	Olds Motor Works	Fire (March 90) at Olds Motor Works; only one curved-dash Oldsmobile saved. After fire, started system of sub-contracting to outside suppliers, setting pattern for the industry.
	Henry Ford Co.	Predecessor company to Cadillac Motor Car Co. organized (November 30)
	Packard	Steering wheel replaced tiller (right-hand drive)
1902	Haynes	Chrome-nickel steel in automobile
	Autocar	Shaft drive to rear wheels via U-joints
	Locomobile	First 4-cyl. production engines (Locomobile)
1903	Buick	Buick Motor Co. organized in Detroit (May 5)
	Ford Motor Company	Incorporated in Detroit (June 16)
	Clyde Coleman	Electric starter patent # 745,157 Nov. 24
	Winton	First 8-in-1ine engine, Winton Bullet #2 Racing Car
	Premier Wiedley	Air-cooled eight-in-line overhead-valve racing engine
	A. 0. Smith	First pressed steel frame
	Peerless	Tilting steering columns
	A. W. Spicer	Universal joint and propeller shaft replaced chain drive
	Tincher	Air brakes
		Glass windshields introduced
1904	SAE Jan. 19	Formation of Society of Automobile Engineers (forerunner of SAE), New York City: first president. Andrew L. Riker
	Sizaire-Naudin (France)	Independent front suspension
	Pierce-Arrow	Engine full-pressure lubrication
	Holley, Shebler, Detroit Lubricator, etc.	Automatic carburetors
	E. V. Hartford	Shock absorbers
	Presto-Lite	Acetylene gas for automobile lamps (Packard)
1905	Remy Brothers	First American magneto for ignition
	National	First 6-cyl. engine
	Industry	Folding tops introduced
	Charles Y. Knight	Sleeve valve engine patented; successfully applied, 1908
	Reo	Radiator had flattened tubes, able to expand in case of overheating or freezing
1906	Ray Harroun	Spring front bumper developed
	Moon	Left hand drive (steering wheel)
	Dr. Foettinger (Germany)	Invented torque converter
	Haynes	Transmission had free-wheeling feature
	Maxwell	Constructed a 12-cyl. racing car
	T. J. Sturtevant (U.S.)	Designed and built a 4-cyl. experimental car with leverless control at Sturtevant Mill Co. of Boston. A three-forward speed (automatic) transmission with gears engaged by centrifugal clutches. He also designed the Sturtevant power-actuated brake, the first known use of the "servo-system."
	Fred Marriot-Stanley	Drove Stanley Steamer a mile in 28.2 seconds, avg. speed 127.66 mph, Ormond Beach, Florida
	Northern	Airbrakes and air clutch control on Northern passenger cars
1907	Clyde J. Coleman	Second patent, electric starter, #842,827 Jan. 29
	Hewitt Motor Co.	Introduced a water-cooled V-8 engine-powered limousine
	Marmon	Built air-cooled V-8 engine
1908	Thomas Flyer	Wins New York-to-Paris race
	Cadillac, Henry M. Leland	Received DeWar Trophy for interchangeability of parts
	C. Harold Wills	Developed use of vanadium steel for Ford Motor Co.
	Otto Zachow & William Besserdich-FWD	Developed first successful 4-wheel-drive vehicle (Clintonville, Wisconsin). Production, 1910.
	Morse-Whitney/Link Belt	Silent camshaft timing chains
	Chadwick	First engine supercharger

In a tough game, you need a good partner.

Pick our brains.

GE Automotive

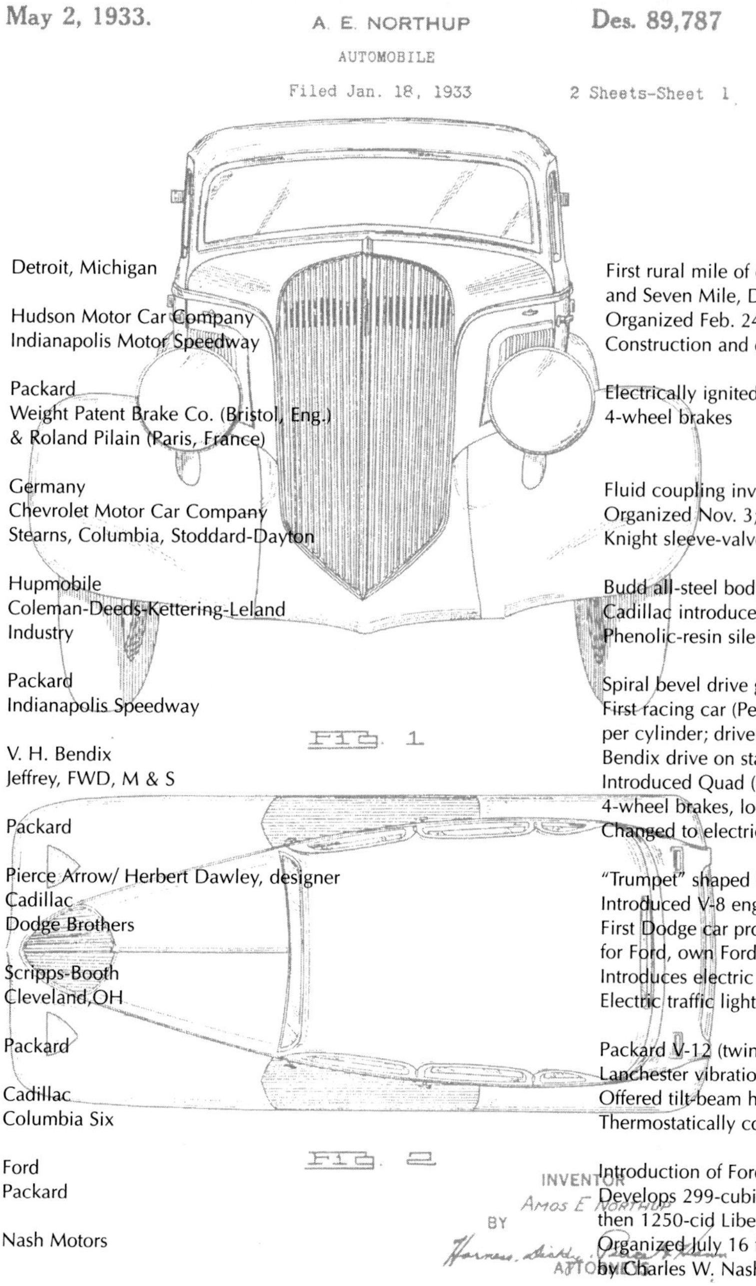

1909	Detroit, Michigan	First rural mile of concrete pavement (Woodward between Six and Seven Mile, Detroit) July 4
	Hudson Motor Car Company	Organized Feb. 24; first car produced July 3
	Indianapolis Motor Speedway	Construction and opening of Indianapolis Motor Speedway
1910	Packard	Electrically ignited acetylene headlamps
	Weight Patent Brake Co. (Bristol, Eng.) & Roland Pilain (Paris, France)	4-wheel brakes
1911	Germany	Fluid coupling invented
	Chevrolet Motor Car Company	Organized Nov. 3; first car tested by Louis Chevrolet
	Stearns, Columbia, Stoddard-Dayton	Knight sleeve-valve engine introduced on Stearns.
1912	Hupmobile	Budd all-steel body on Oakland and Hupmobile
	Coleman-Deeds-Kettering-Leland	Cadillac introduces electric starter
	Industry	Phenolic-resin silent camshaft timing gears
1913	Packard	Spiral bevel drive gears in rear axle
	Indianapolis Speedway	First racing car (Peugeot) with overhead camshaft, 4 valves per cylinder; driver Jules Goux
	V. H. Bendix	Bendix drive on starters on Chevrolets
	Jeffrey, FWD, M & S	Introduced Quad (4-wheel drive) truck, 4-wheel steer & 4-wheel brakes, locking differential
	Packard	Changed to electric headlamp lighting
1914	Pierce Arrow/ Herbert Dawley, designer	"Trumpet" shaped headlamps in front fenders
	Cadillac	Introduced V-8 engine
	Dodge Brothers	First Dodge car produced (Nov. 14); continued to produce parts for Ford, own Ford stock
	Scripps-Booth	Introduces electric door latches
	Cleveland,OH	Electric traffic lights installed
1915	Packard	Packard V-12 (twin six) engine with aluminum pistons and Lanchester vibration damper
	Cadillac	Offered tilt-beam headlights
	Columbia Six	Thermostatically controlled radiator shutters
1916	Ford	Introduction of Fordson tractor
	Packard	Develops 299-cubic-inch V-12 racing engine; later, 905-cid, then 1250-cid Liberty aircraft engine
	Nash Motors	Organized July 16 from Thomas B. Jeffrey Co. by Charles W. Nash, former GM president
1917	Budd, Michelin	Steel disc wheels
	Chevrolet	Chevrolet V-8 engine
	Packard	Packard V-12 sets speed records at Sheepshead Bay (130.4 mph)
	Studebaker	"hot spot" manifold to improve fuel vaporization
	Paige-Franklin	V-type windshield
	Nash car	Nameplate first applied
	Moon	Stylish one-piece plate-glass windshield
1918	Nash	Becomes world's largest producer of trucks (11,494)
	Nash 4-pass. coupe	First pillarless coupe body design
1919	Packard, Franklin	Electric vaporizer in intake manifold
	Ultra Safety Glass	Shatterproof safety glass (first applied, 1927) Glass Founders Corp.,Milltown, NJ
1920	Geo L. Weiss, W. D. Packard Engr Co.	"Lovejoy" hydraulic shock absorber
	Packard	Lanchester vibration damper on single six
1921	Dr. Thomas Midgley, Jr.	Develops tetra-ethyl lead additive for gasoline, Dayton Engineering Labs
	Duesenberg	America's first 8-in-row, overhead-camshaft 8-cyl. engine, with 4-wheel hydraulic brakes
1922	Lincoln Motor Co.	Acquired from Henry Leland by Ford Motor Co. (Feb. 4)

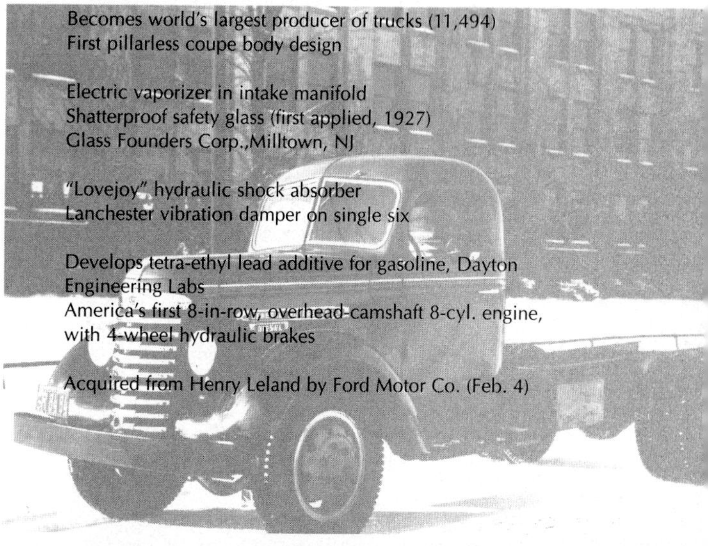

FIG. 1

FIG. 2

INVENTOR
Amos E. Northup
BY
ATTORNEY

1923	Aluminum Manufacturing Company	Cast aluminum disc road wheel
	Rickenbacker Motor Car Co.	Tandem flywheel
	Cole V-8	Balloon tires (with Firestone), clay modeling
	Oakland (GM), Cole, DuPont Chemical	"Duco" nitro-cellulose fast-drying lacquer finish
	Packard	Introduces in-line-8 engine with Lanchester vibration damper, 9 main bearings and 4-wheel mechanical brakes
1924	Chrysler	Introduces (January) high-compression, 7-main-bearing, 6-cyl. car with hydraulic 4-wheel brakes, Purolator crankcase oil filter
	Chandler	Constant-mesh "Traffic transmission;" forerunner of synchromesh transmission
	GM-Ethyl Gasoline Corporation	Distribution of ethyl high-octane gasoline
1925	Industry	Balloon tires standard on most cars
	Chrysler Corporation	Incorporated (June 6)
	Packard	Skinner lubricating oil rectifying system; forerunner of full-flow filter
1926	Packard	Hypoid rear axle (angle set)
	Stutz, Rickenbacker	Safety shatterproof glass in production cars
	Chrysler Finer "70"	18-inch wheels, 30 x 6.00 tires
	Stutz	Underslung worm-drive rear axle
	Pierce-Arrow	Bragg-Kliesrath power-brake booster

1927	Reo Flying Cloud	Internal-expanding 4-wheel hydraulic brakes
	Paige Detroit	4-speed internal gear transmission
	Reo Wolverine	Full crankcase ventilating system
	Studebaker	Westinghouse Air power brakes
	Oldsmobile, Studebaker	Chromium plating of bright metal
	AC Division, GM	Mechanical fuel pump
1928	Dodge Victory Six	Body floor-panel integral with car frame; forerunner of unitized body
	Great Britain	Fluid coupling on London buses
	Ford, Owen Dyneto	Electric windshield wipers
	Ford Motor Co.	Triplex shatterproof windshield on Ford, Lincoln
	Plymouth	First produced (June 14)
	Dodge	Acquired by Chrysler Corp. (July 30)
1929	Cadillac	Selective synchromesh transmission
	Cord	Front-wheel-drive
1930	Chrysler	Downdraft carburetor
	Nash	Starter button on instrument panel
	Cadillac	V-16 overhead-valve engines
	Oakland, Viking	V-8 engine in low-priced car
1931	Marmon	V-16 engines
	Plymouth & Studebaker	Vacuum spark retard and centrifugal spark advance
	Industry	Freewheeling on Auburn, Plymouth, Studebaker, Graham, Hudson, Hupmobile, Marmon, Peerless, Willys
	Industry	Startix (Bendix) automatic starting on Auburn, Hudson-Essex, Rockne, Studebaker
	Oldsmobile	Synchromesh transmission
	Reo Royale	"Mother of pearl" lacquer finish
	Chrysler	Convex (spherical) headlamp lens (C. M. Hall Lamp)
	Budd	Centrifuge brake drums on Chrysler & Imperial
	Auburn	X-member chassis frame

1932	Duesenberg	Supercharger on Duesenberg SJ
	Packard, Pierce-Arrow	V-12 engines with hydraulic valve silencers
	Industry	Silencing improvements in fan, muffler, air intake
	Reo Royale	Longest wheelbase 7 passenger sedan 152 inches
	Goodyear Airwheel Tires	7.50 x 15 Extra Low Pressure Tires
	Packard, Olds	Automatic carburetor choke
	General Motors & Chrysler Corp	Vacuum operated clutches — Buick, Cadillac, Chrysler, DeSoto, Dodge, Plymouth
	Chrysler Imperial	Stellite exhaust valve seats
	Packard, Imperial, Duesenberg	Power brakes

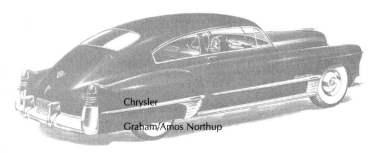

Chrysler
Graham/Amos Northup

Year	Make/Model	Feature
		Shell-type main and rod bearings, Needle roller bearings in propellor shaft U-joints
		Bold new trend in styling: "Blue Streak" with sloping windshield pillars, omission of traditional radiator shell; radiator filler under the hood; clam-shell-type fenders with wheel opening following tire outline; chassis had banjo-type frame at the kick-up location, with parabolic openings in frame-rail web for rear axle tube; frame had outrigger spring hangers and body brackets; engine featured "Bohnalite" aluminum head
1933	Reo Royale	2-speed self-shifting transmission (no clutch pedal); forerunner of the automatic transmission
	Pierce Silver Arrow	Pressed metal roof on closed cars
	Franklin	"Nl Resist" exhaust valve seats
	Chevrolet & Pontiac	Flanged rear axle shafts
1934	Chrysler, DeSoto	Warner overdrive on Chrysler & DeSoto Airflow
	Chrysler & General Motors	Independent front wheel suspension on production cars, (knee action) — GM, Plymouth, Dodge, Chrysler Six
	Chrysler, DeSoto	Semi-unitized body construction, Airflow models
	Chrysler	One-piece, curved glass windshield, Custom Imperial Airflow
	Ford	Clear vision ventilation
	Studebaker	Hill holder on transmission (November)
	Hudson, Terraplane	Pre-selected electro-vacuum gearshift
		One piece stamped steel roof on closed models
1935	GM	All steel turret top, pressed metal roof in production
	Packard, Stout	Independent suspension with torque arm
	Packard, Chrysler	Raised body belt-line and chair-height seats
	Nash	Unitized body construction
1936	Studebaker	"Safety Catch" rotary door latches
	Cord 810	Concealed headlamps, front-wheel-drive V-8; pre-selected electro-vacuum gearshift; unitized construction (subframe with powertrain module bolted to the unitized body)
	Stout Scarab	Rear-engined, open-seating "minivan" predecessor; only a handful produced
1937	Chrysler	Adjustable front seat—fore/aft, up/down
	Chrysler Corp.	Safety flush-faced instrument panel and controls
	Graham	Graham-Bradley farm tractor for Montgomery Ward distribution
1938	Buick, Olds	Four-coil spring suspension
	Nash	Full-sized bed optional on sedan models
	Pontiac	Column-mounted gearshift lever
	Nash	"Weather Eye" thermostatically-controlled heating, air system
1939	Olds	Hydramatic drive, 1939 Oldsmobile
	Chrysler	Fluid drive, fluid coupling, 1939 New Yorker & Imperial
	Hudson	Airfoam cushions
	Packard	Air conditioning in passenger cars
	Chrysler, Dodge, DeSoto	Pillarless 5-passenger club coupe
	Chrysler	2-speed electric windshield wiper motors
	GM & Chrysler	Sliding sunroof
1940	Industry	Sealed beam headlamps on 1940 models; most cars have column shift
	American Bantam	General purpose (GP) 1/4 ton vehicle developed for Army; Willys and Ford contracted to build; later nicknamed "Jeep" from the initials GP
	Nash "600"	"Unitized" body-and-frame construction
1941	Packard Clipper	Innovative styling (fade-away fenders)
	Packard	Elimination of bolt-on rear fenders
	Packard, Cadillac, Imperial	Electro-hydraulic power windows and power seat
	Chrysler	"Safety rim" wheels on 1941 models
	Buick	Dual carburetors

1942	Industry	Passenger car production suspended for duration of WWII (Ford produces last civilian car Feb. 10)
1946	Willys Jeep	Civilian versions of WWII military vehicle plus all-steel station wagon model
	Crosley	Overhead camshaft engine in a "minicar"
1947	Studebaker	First post-war models have "two-way" styling with wrap-around backlite
1948	Tucker	Butyl-bonded windshields; about 50 produced
	Cadillac	Rear body-quarter fins
	Buick	"Dynaflow" torque-converter automatic transmission
	Hudson	New post-war "step-down" body design
	Industry	"Low-pressure" tires for better ride
1949	Kaiser, Frazer,	Hatchback body design (Kaiser Traveler & Frazer Vagabond)
	Chrysler Corp. cars	Ignition-key starting control
	Cadillac, Oldsmobile	High-compression overhead valve V-8; Cadillac's "oversquare" 331-cid at 150 hp @ 3800 rpm with hydraulic valve lifters, five main bearings and 7.5:1 compression ratio replaced a flathead 140-hp @ 3400 rpm 346-cid engine with three mains and 7:25 compression ratio. Olds "Rocket V-8" was 303 cid and 135 hp, founded "88" series with later "stock car" auto-racing successes.
	Nash	Curved one-piece safety glass windshield
	Triumph (England)	Coilspring telescope-strut damper unit
1950	Ford of England	McPherson-strut independent front suspension
	Chrysler	4-wheel disc brakes on limited-production Crown Imperial
	Nash	Seat belts in Rambler model
	Chevrolet	First automatic transmission in "low price" field, "Powerglide" torque-converter type
1951	Chrysler	"Fire-power" hemispherical combustion chamber engine on New Yorker & Imperial
	Nash-Healy	Headlamps located in front grille
	Buick	Tinted glass
	Buick, Chrysler	Power steering
1952	Cadillac	Four-venturi and four-throttle-blade carburetor
	Packard	"Easamatic" treadle-vac power brakes
	Ford Motor Co.	Suspended brake and clutch pedals
	Chrysler	12-volt electrical system, Crown Imperial
1953	Packard	Orlon convertible top material introduced on Caribbean
	Chevrolet Corvette	Sports car with two-place, glass-fiber body, floor-shift gear selector
1954	Packard	Tubeless tires (by B. F. Goodrich)
	Ford, Mercury	Transparent plastic body roof insert
1955	Packard	Packard torsion-level ride, full-length torsion-bar suspension with electrical automatic leveling
	Thunderbird	Removable hardtop roof
	Packard	Nylon and Orlon interior trim materials
1956	Cadillac, Packard, DeSoto, Plymouth	Color anodized aluminum exterior trim
	Packard, Chrysler Corp.	Push-button automatic transmission controls
	Chrysler Imperial	Electrically-controlled door latches
	Ford	"Safety package" option with seatbelts, extra padding
1957	Chevrolet, Rambler	Fuel injection (vacuum-mechanical and electronic)
	Ford Ranchero	Passenger-car-styled pickup truck
	Ford Skyliner	Retractable hardtop convertible
	Thunderbird	Telescoping steering column
	Chrysler, GM	Viscous fan drive

	Imperial	Curved door glass and compound-curved windshield
	Mercury Turnpike Cruiser	Windshield-header-mounted fresh-air intakes
		Power-operated retractable rear window
		Power-operated front seat mechanism with memory control
	Imperial, Chrysler, Mercury	Four 5.75-inch-diameter "dual" headlamps
	Plymouth	Rear-facing third seat in station wagon
1958	GM, Mercury, Imperial	Air suspension
	Chrysler & Imperial	Cruise Control
	Edsel	Self-adjusting brakes
	American Motors	Re-introduction of 100-inch-wheelbase "American" model as "compact car" — model had been discontinued in 1954
1959	Chrysler Corp.	Swivel front seats
	Morris Minor (England)	Front-wheel-drive with transverse engine
	Various	Outside rearview mirror adjustable from inside
1960	Plymouth Valiant	Alternator charging system as standard equipment)
	Chevrolet Corvair	Air-cooled opposed six-cylinder engine in rear with trans-axle drive
	Ford Falcon, Valiant, Corvair	Introduction of "compact" passenger cars by Big Three to supplement "full-sized" Ford, Plymouth, Chevrolet
	GMC Truck	V-6 and V-12 truck engines
	NSU (Germany)	Rotary engine developed
1961	Ford Motor Co., Cadillac	Extended chassis lubrication intervals
	Ford Econoline, Chevrolet Greenbrier	"Box on wheels" light truck designs with forward control, chassis adapted from compact Ford and Chevrolet passenger cars; patterned after European models
	Buick Special, Oldsmobile F-85	Aluminum V-8 block (215 cid)
	Cadillac & AMC	Dual master-brake-cylinder
	Lincoln Continental	Return of four-door convertible
	Pontiac Tempest	"Half-eight" slant four cyl. with flexible driveshaft and rear transaxle
1962	Goodyear	Polyester fibre tire cord
	Buick Special	First U.S. passenger car V-6 engine
	Ford Fairlane, Mercury Meteor	"Intermediate-sized" cars introduced; product proliferation concept begins
1963	Goodyear	Tire "inner spare"
	Pontiac	Transistorized ignition system (option)
	Industry	Positive crankcase ventilation system for emission control Seatbelt anchors standard for outboard front-seat occupants
	Chrysler Corp.	Turbine-engine powered car (limited production)
1964	Industry	Front seatbelts standard
	Ford Mustang, Plymouth Barracuda	Sporty "pony cars" introduced, based on compact chassis
	Cadillac	Automatic headlamp on-off control
1965	Industry	Automatic transmission quadrant uniform with "P-R-N-D-L" pattern; rearseat belt anchors standard
1966	Oldsmobile Toronado	Re-introduction of front-wheel drive
	Ford Bronco	4-wheel-drive utility/recreation vehicle by major manufacturer;
	Pontiac Tempest	Overhead camshaft 6-cyl standard
	Lincoln Continental	Power front disc brakes standard
	GM, Chrysler, AMC	Collapsible steering column
	Industry	Rear seatbelts standard
1967	Industry	Dual braking systems standard
	GM	Ignition-key-in-switch warning buzzer
1968	Industry	Exhaust emission controls standard; shoulder belts mandatory for outboard front seat occupants
	Ford Motor Co.	"Controlled-crush" front body design (selected models)
	GM	Ventless side windows
	American Motors	Headlamps on warning buzzer

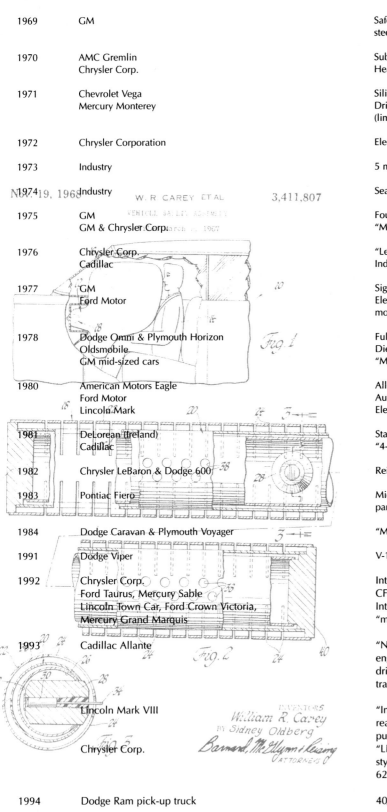

1969	GM	Safety door beams for side impacts; reintroduction of locking steering columns as anti-theft measure
1970	AMC Gremlin Chrysler Corp.	Sub-compact Headlamp delay system
1971	Chevrolet Vega Mercury Monterey	Silicon-impregnated aluminum cylinder block Driver "air bag" supplement for front impact protection (limited production for selected fleets)
1972	Chrysler Corporation	Electronic ignition
1973	Industry	5 mph crash front bumpers
1974	Industry	Seatbelt ignition interlock system (discontinued after 6 mo.)
1975	GM GM & Chrysler Corp.	Four rectangular head lamps (selected models) "Monolithic" catalytic converters for emission control
1976	Chrysler Corp. Cadillac	"Lean Burn" engine to control emissions Industry's last production convertible — for several years
1977	GM Ford Motor	Significant downsizing and weight reduction of full-sized cars Electronic anti-lock brakes (standard on selected low volume models — Lincoln Versailles, Mercury Grand Monarch)
1978	Dodge Omni & Plymouth Horizon Oldsmobile GM mid-sized cars	Full production of domestic front-drive transverse-engine cars Diesel V-8 (converted from 350-cid gasoline engine) "Mini-spare" short-range temporary spare tire
1980	American Motors Eagle Ford Motor Lincoln Mark	All-wheel-drive passenger vehicle Automatic overdrive transmission (selected models) Electronic instrument cluster
1981	DeLorean (Ireland) Cadillac	Stainless-steel body sports car "4-6-8" V-8 engine option (discontinued after one year)
1982	Chrysler LeBaron & Dodge 600	Reintroduction of folding top convertible
1983	Pontiac Fiero	Mid-engine, two passenger car sports car, "Enduroflex" plastic body panels
1984	Dodge Caravan & Plymouth Voyager	"Minivan" garageable small, forward-control passenger van
1991	Dodge Viper	V-10 engine in sports car
1992	Chrysler Corp. Ford Taurus, Mercury Sable Lincoln Town Car, Ford Crown Victoria, Mercury Grand Marquis	Integrated child restraint seat. CFC-free air-conditioning. Introduced 4.6-liter SOHC iron-block V-8, first of new "modular" engine family.
1993	Cadillac Allante	"Northstar" 32-valve, 290-hp, 4.6-liter V-8 aluminum block engine w/quad overhead camshafts, mated to front-wheel drive electronically controlled 4-speed automatic transmission. (April 23, 1992 public introduction)
	Lincoln Mark VIII	"InTech" 32-valve, DOHC 280-hp, 4.6-liter aluminum V-8 rear-drive version of "modular" engines. (Dec. 26, 1992, public introduction).
	Chrysler Corp.	"LH" series six-passenger sedan family with "cab forward" styling, longitudinally-mounted V-6 front wheel drive, wide 62-inch stance.
1994	Dodge Ram pick-up truck	40/20/40 split front seat and V-10 engine option.
1995	Lincoln Continental, Ford Contour, Mercury Mystique	"MicronAir" filtration system.

Independent Automakers

Quaint-sounding now, firms weren't marginal, but many are only memories

by Anthony J. Yanik

TOP: *Baseball great Babe Ruth poses with 1937 Nash Ambassador 8. (Chrysler Historical Collection)*

The term "independents" materialized in the latter half of the 1930s. By then, the Great Depression was on the wane. Auto companies that had survived but were not members of the Big Three (General Motors, Chrysler, and Ford) found themselves relegated to this hypothetical category.

The Independents consisted predominantly of the Hudson Motor Car Company, Nash Motors, the Packard Motor Car Company, Studebaker Corporation, and Willys-Overland. They were the highest volume producers outside of the Big Three although their combined sales represented slightly less than 10 percent of the market. Still, they are worthy of mention because of their long, rich blood lines, lines that extended into the very early days of the automobile.

Most revered was the Packard Motor Car Company. Packard was the oldest of the Independents, having been officially incorporated in Warren, Ohio in 1900. By 1903, wealthy Detroiter Henry Joy had purchased a controlling interest in Packard with his friends, and moved the company to Detroit. Joy's goal was to produce one of the finest luxury cars in the United States. For the next four decades he succeeded admirably. Outside of Cadillac, which benefited from General Motors' protection, and Lincoln, which was supported by the Ford Motor Company, Packard was the only luxury car manufacturer to survive the Great Depression — and only by a whisker. The company revived after World War II and experienced its best sales ever in 1948. It

was a short-lived success. Six years later Packard was forced to merge with Studebaker in order to survive, but the merger was ill fated. By 1958 the Packard marque was dropped.

Second oldest of the Independents was Willys-Overland, founded as Overland in 1903 in Terre Haute, Indiana. The company almost expired during the panic of 1907, but was rescued by one of its dealers, John North Willys. Willys seized control of the virtually defunct company and made it solvent. By 1912, Willys had become the number two seller behind Ford. It held this position until the severe recession that followed World War I forced the company to borrow a large sum of money from the Chase National Bank in Manhattan to meet its debts. Chase insisted that Wal-

Roy D. Chapin, Sr., with 1933 Hudson roadster. (Chrysler Historical Collection) BELOW: John North Willys. (National Automotive History Collection, Detroit Public Library) OPPOSITE, TOP: 1946 Willys Jeep station wagon, one of the most significant automotive concepts of the century, forerunner of the highly popular sports/utility vehicles offered by most automakers in the 1990s. (Chrysler Historical Collection) OPPOSITE, BELOW: George W. Romney, CEO of American Motors in the late 1950s and early 1960s before turning to public life as governor of Michigan. He is credited with being father of the "compact car" concept. (Chrysler Historical Collection)

ter P. Chrysler be brought in to manage Willys-Overland's affairs. Chrysler remained for two years. When he left, the company was solvent again, and John North Willys had maneuvered himself back into the company's chairmanship.

BRIEF RESURGENCE

During the 1920s, Willys-Overland once more became a significant force within the industry, but the Great Depression sent it reeling into receivership. The company was reorganized, and survived, but no longer as a major player. Its place in auto history has been secured by virtue of its being prime builder of the indomitable Jeep of World War II fame. Willys was absorbed by the Henry J. Kaiser Company in 1953 and renamed Willys Motors, Inc. The passenger car portion of Willys was abandoned in mid-1955 so that Kaiser could focus its energies upon the commercial vehicle Jeep line. The latter became so successful an operation that Kaiser changed the name of Willys Motors, Inc. to the Kaiser Jeep Corporation in March 1963. Ownership of Kaiser Jeep changed hands in February 1970 when the company was purchased by American Motors. On Aug. 5, 1987, American Motors Corporation was acquired by

Chrysler Corporation primarily for the highly successful Jeep line. Thus Willys-Overland continues to live within the Jeep nameplate.

The Studebaker brothers were the leading wagon makers in the nation when they decided to dabble in automobiles. Their first entry was an electric designed by Thomas Edison in 1902 followed by a gasoline powered vehicle in 1903. Neither was marketed very seriously. In 1908, Studebaker purchased a one third interest in the stock of the newly formed E-M-F (Everitt-Metzger-Flanders) Company, buying out that concern in 1912 after E-M-F rose to third in domestic sales. By the mid-1920s, only Ford and General Motors outranked Studebaker in total assets. A series of poor business decisions forced the company into receivership in 1933, but it revived under new management. Studebaker remained moderately successful for the next 20 years, then saw its fortunes decline along with those of the other independents. It merged with Packard in 1954, a move that prolonged its life for another dozen years. The last Studebaker car was built in 1966.

"Closed Cars' Opening

The Hudson Motor Car Company was formed in 1910 by Howard E. Coffin and Roy D. Chapin. Joseph L. Hudson of department store fame supplied its capital. It was Hudson's Essex Coach of 1921 that created an industry trend toward closed cars. Until the Essex came along, 90 percent of cars sold were open models because they were 30 to 50 percent cheaper than closed cars. Hudson's owners cut the cost of manufacturing costs of a closed car drastically which enabled them to offer the Essex Coach at a price slightly above that of an open car. So popular was public response to a low priced closed car that other makers soon followed Hudson's leads. Hudson survived the Great Depression, and prospered in the sellers' market that followed World War II, but sales slipped 50 percent between 1949 and 1953. Its merger with Nash in 1954 did not bear fruit. The Hudson nameplate disappeared from active ranks in 1957.

Fifth member of the Independents group was the Nash Motor Company owned by Charles W. Nash. Nash had worked his way up from iron sorter in the Durant-Dort Carriage Works to the presidency of Buick, then General Motors. He resigned from GM in 1916 when Durant regained control of that company, and purchased the Thomas B. Jeffery Company, changing the latter's name to Nash. For the next two decades, Nash cars were among the top ten in sales. In 1936, at the age of 72, Nash retired from active management. He selected George Mason of Kelvinator, maker of refrigerators, as his successor, resulting in a Nash-Kelvinator merger. Mason later was joined by George Romney who became president in 1954 when Mason died. Nash followed the trend of other Independents and merged with Hudson in 1954 to form the American Motors Corporation (AMC). Romney's emphasis on building small cars brought success to AMC, so much so that

through the 1960s and most of the 1970s, the Big Three often was referred to as the Big Four. In 1970 AMC felt strong enough to take over Kaiser-Jeep. During the later 1970s, however, as the sales of imported small cars rose markedly, AMC fortunes declined. In 1979, Renault of France purchased 46.6 percent of AMC Stock, effectively taking control of the company. By

marked the beginning of the company's death throes, and no Hupmobiles were built in 1937. In 1938 the company resorted to the unusual move of buying the dies of the front wheel drive, coffin-nosed 810/812/ 1936/1937 Cord and reworking them into a rear drive Hupmobile offshoot. Hupmobile's existence came to an ignoble end in 1939 when the City of Detroit refused to provide it relief from delinquent taxes and the Hupp Motor Car Company went bust paying them. The last car carrying the Hupmobile name was built in July 1940.

The Graham-Paige Motor Corporation originally was formed by Harry Jewett in 1909 as the Paige-Detroit Motor Car Company, Fred Paige being its president. Sales were mediocre at best until 1923 when they reached a high of 43,556 units. Jewett thereafter tired of the automobile wars, and accepted an offer from the brothers Graham in 1927 to buy him

1987, Renault, now suffering money problems of its own, accepted an offer from Chrysler Corporation to purchase AMC. The latter subsequently became the Jeep-Eagle Division of Chrysler.

A second tier of "Independents" bearing mention would be the Hupp Motor Car Company, Graham-Paige, and Kaiser-Frazer.

ONE-HUPPMANSHIP

The Hupp Motor Car Company was organized on Nov. 8, 1908. Its first car, a four-cylinder runabout called the Hupmobile, was introduced in February 1909 at the Detroit Automobile Show. It was priced at $750. Two years later,

One of the lesser independents in Detroit in the 1920s and 30s was Graham-Paige. This "shark-nosed" Supercharged Graham of 1938 won gas economy trials and art deco styling acclaim, but like Chrysler's Airflows of the middle '30s, failed to capture the conservative public taste. The company survived after World War II as Kaiser-Frazer. (National Automotive History Collection, Detroit Public Library) BELOW: Packard, more than any other make, always symbolized the classic car. This ca. 1930 classic Packard shown here was photographed at a meet at Greenfield Village in 1958. (Mike Davis collection)

Robert Hupp resigned after a row with other company officers who desired to add an expensive touring car design to accompany the low priced runabout. (Hupp had once been a car tester for R.E. Olds. Like Olds he had been forced out of his own company, and had gone on to form another under his initials. It went bankrupt in two years, and Hupp subsequently disappeared from the automotive scene.)

The Hupp Motor Car Company (minus Robert Hupp) grew slowly, and annual sales never exceeded the five figure mark. Hupmobile's best year was 1928 when it built 65,862 cars. The Great Depression

out. The Grahams had sold their highly successful truck manufacturing company to Dodge Brothers the year before, but found themselves with too much time on their hands, hence the Paige-Detroit purchase. They changed the company's name to the Graham-Paige Motors Corporation, and introduced an entirely new line of automobiles.

Their first year production of 73,195 cars was the highest ever recorded by a new line to that date. Plans to expand production were immediately considered, but cast aside once the Great Depression began. In 1932, Graham-Paige sales plummeted to 12,967. The

Grahams struggled to keep the company solvent, and succeeded.

In August 1944, Joseph Graham sold a large block of the company's stock to Joseph W. Frazer, former head of Willys-Overland. Immediately thereafter Frazer was voted in as chairman of the board. Frazer hired Howard Darrin to design a radically new car with the intent of having a first rate product available for sale in the buyer's market that was certain to follow the end of World War II. To carry out these plans, he needed more money.

TWO-FER-ONE PRODUCTION

Enter the short-lived Kaiser-Frazer Corporation with Joe Frazer as its president. Frazer also would continue as president of Graham-Paige, and the two companies would build cars off the same Darrin design sharing the same plant! They then purchased the empty Willow Run ex-bomber plant, and began to put into place their unique form of two-company production. Output was to be two Kaiser cars for each Graham-Paige, now called the Frazer. Costs were to be divided on the same ratio: two-thirds Kaiser, one-third Graham-Paige.

When Kaiser and Frazer prototypes were displayed at the Waldorf-Astoria in New York in January 1946, they created a sensation, More investment money was needed. Kaiser quickly raised his two-thirds share, but Frazer failed. Kaiser then proposed that Frazer sell the Graham-Paige automobile operations to Kaiser-Frazer so that he could have a free hand in raising the necessary finances. The sale took place on Feb. 5, 1947. Exit Graham-Paige as an automobile producer.

Kaiser-Frazer now stood on its own merits. Initially it did very well. Sales exceeded 100,000 units each year between 1947 through 1951, elevating Kaiser-Frazer among the top ranks of the Independents. A diminutive Henry J. Model was added, and a drastic redesign of the Kaiser made in 1951. The company felt strong enough to purchase Willys-Overland in April 1953 for the rights to the latter's commercial Jeep

vehicles, but Kaiser sales had fallen off badly. Car production finally was halted in June 1954. The next month the Kaiser assembly lines were switched exclusively over to Jeep commercial vehicles.

Today the term "Independents", is rarely used, if ever. The only companies that might qualify would be U.S. transplants of such worldwide corporations as Honda, Nissan, Toyota, Mazda, Mitsubishi and BMW.

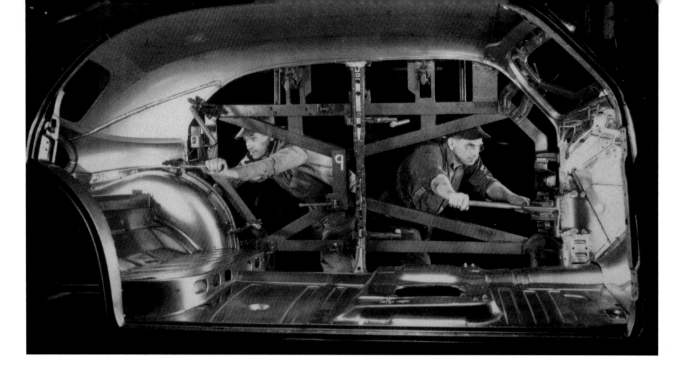

Fisher Body, organized in 1908 by a family of buggy makers to supply bodies for the rapidly growing automobile industry, was acquired by General Motors in the 1920s. GM kept the Fisher Body plants separate from car assembly plants until the 1980s. Here, in about 1941, Fisher body workers install temporary bracing required to position the door pillar before welding it into place. In 1995, such tasks are handled by robots. (General Motors)

Suppliers Meet Demand

Relying on outside skill demonstrated automakers' savvy

by Al Fleming

It can be argued that the automotive supplier industry emerged from a fire that destroyed Detroit's Olds Motor Works plant in March 1901. The only thing saved was an experimental curved-dash roadster. In getting back to business as fast as possible following the disaster, Ransom E. Olds had no choice but to use the one surviving car as a model to order parts and subassemblies from small shops in the Detroit area. The curved-dash Oldsmobile runabout became an instant success, encouraging other pioneer auto companies to seek outside sources for parts. Many of those shops became major automotive suppliers, some became auto manufacturers and Detroit became the Motor City.

So the theory goes. But there were other factors behind the origin and evolution of Detroit's — and America's — automotive supplier industry.

From the beginning at the turn of the century, automobile companies were primarily parts assemblers, not manufacturers. They relied on outside firms for vehicle bodies, engines, wheels and other components. The reason was purely economic. The cheapest, fastest, least risky way to become an auto company was to place the burden of fixed and working capital on suppliers. It made more sense to rent a small assembly plant than to buy one. Besides, vehicles could be assembled faster than the 30- to 90-day credit period parts manufacturers allowed.

So the choice was clear: Once the basic design of a car was established, parts making was jobbed out to independent shops, minimizing the auto company's capital needs for wages, materials, machinery and factory space.

Consequently, the automobile of the early 1900s became a combination of components which were already being produced for other uses such as stationary and marine gasoline engines, carriage bodies and bicycle wheels.

Sources of materials to make parts were also readily available. By the start of the 20th century, there were four rubber companies — Firestone, Goodrich, Goodyear and U.S. Rubber. Dow Chemical was an automotive supplier almost from its inception in 1897 with the discovery of magnesium. In 1901 United States Steel was created from several smaller steel companies.

Little wonder that Ford Motor Co. was able to begin business in 1903 with meager paid-in capital of only $28,000, a dozen workers and a small plant just 50 by 250 feet. The company manufactured nothing, relying on suppliers for vital parts including engines, transmissions and axles purchased from a Detroit machine shop run by brothers John and Horace Dodge.

A Selling Point

Some auto companies even bragged that all they did was assemble parts. Wayne Automobile Co., a Detroit firm, said in 1904, "This gives (us) the advantages of the very best facilities of the best American constructors, far better than any one factory could hope to maintain, and reduces the flat cost to the lowest terms."

Olds Motor Works (later Olds Motor Vehicle Co.) is an example of how "automakers" of that period relied on a wide range of outside suppliers. In 1900, as his company began production, Ransom Olds bought bodies from Detroit's C.R. Wilson Body Co., a carriage maker that had started building bodies for horseless carriages. As production increased, Olds also bought bodies from Byron F. Everitt of Detroit and H. Jay Hayes of Cleveland.

Olds initially ordered wheels from Weston-Mott Co. of Utica, N.Y., a manufacturer of bicycle wheels which started shifting its emphasis to automobile wheels in the late 1890s. By 1902 Olds was purchasing most of his wheels from Prudden Co. in Lansing, batteries from Sipe & Sigler in Cleveland and roller bearings from Hyatt Roller Bearing Co. in Newark, N.J.

Since Olds manufactured engines, he originally had planned to produce other mechanical parts, but demand became so great that by 1901 he decided to have his engines built by the Dodge brothers (who later

gave up the Olds contract in 1903 to supply Ford instead) and Henry M. Leland, whose shop had earlier landed the job of making Oldsmobile transmissions. Drop forgings came from Brooklyn, N.Y., while radiators, fenders and other sheet metal parts came from Detroit's Briscoe Manufacturing Co.

By 1909 Detroit was producing 25 different makes of cars, which prompted many suppliers to locate plants and businesses in other Michigan communities including Pontiac, Saginaw, Flint, Lansing, Monroe, Port Huron, Grand Rapids and Muskegon.

The fast but recklessly growing industry attracted not only original-equipment suppliers, but also entrepreneurs who saw opportunity and fortune in auto parts and service.

'Orphan' Opportunity

Alfred O. Dunk started the Auto Parts Co. in 1908, buying up the assets — parts inventory, drawings, machinery — of bankrupt companies, with the idea of becoming the replacement parts source for the companies' orphan vehicles as they began to break down. At one time he owned 756 auto companies that had gone out of business.

Another visionary was Robert Simpson, a blacksmith in Litchfield, Mich., who in 1912 converted his shop into a garage and repair business, later supplied parts directly to automakers and eventually started Simpson Industries.

Meantime, spiraling automobile production began causing problems for suppliers who were sometimes hard pressed to keep up with customer demand. The failure of just one supplier to provide a needed part could cause costly vehicle assembly delays. Before long the larger, established auto companies began taking matters into their own hands.

Ford, the volume leader in those days, assured itself of the parts it needed by taking the entire production of some suppliers, making them virtually subsidiaries. In addition, Ford began manufacturing many of its own parts.

Another solution to the problem was to gain control of parts companies. That was the approach taken by General Motors, under William C. Durant, which acquired control of Weston Mott, Hyatt Roller Bearing, Delco, Remy Electric, AC Spark Plug and other suppliers in the years 1909-1919.

Along the way, some of Detroit's pioneer suppliers became multi-millionaires and some became vehicle manufacturers themselves. The Dodge brothers had opened a machine shop to make transmissions for Olds Motor Works in 1901-1902. In 1903 Henry Ford

Dodge touring car bodies, which began production at Budd Manufacturing Co.'s plant in Philadelphia in 1915, were loaded onto freight cars and shipped some 700 miles to the Dodge Brothers plant in Detroit. The all-steel bodies wholesaled for $42 each. (The Budd Co.)

ducers declined steadily after 1920, the number of supplier companies kept increasing.

In the free-wheeling days following World War I, many entrepreneurs became interested in entering the lucrative auto market; but few could dream of doing so if they had to produce their own parts. Fortunately, many reliable manufacturers were available specializing in bodies, engines, transmissions, ignition systems, radiators and brakes — whatever the auto company desired. The supplier business prospered, and automobiles built of a conglomeration of parts by various suppliers became known as "assembled" cars.

The "assembled" car had some advantages. As a rule, it was built of excellent parts made by well-known and respected manufacturers who sought to achieve top quality by specialization.

offered each brother 50 shares of stock in his new company, provided they would produce engines for him. In 1914 the Dodges started up their own auto company, and five years later received $25 million for their original investment in Ford (plus many millions in Ford dividends in the intervening years).

BUDDING INDUSTRY

In 1908 Fred J. and Charles T. Fisher organized Fisher Body Co. In 1919 General Motors bought an interest in the firm (then bought it lock, stock and barrel in 1926 for an estimated $208 million in GM stock), and Nash Motor Co. acquired an interest in Seaman Body Corp. Those moves heralded the virtual disappearance of independent body- and coach-building companies within two decades. An exception was Budd Manufacturing Co., whose first product was the Oakland all steel touring-car body built in Philadelphia starting in 1912. In 1925 the Liberty Motor Car Co. in Detroit went into receivership and Budd acquired the 86-acre plant complex on Charlevoix and Conner in southeast Detroit.

Meantime, predictions that the trend by auto manufacturers toward producing their own parts would wreak havoc on independent suppliers were not borne out. In fact, while the number of motor vehicle pro-

COMPONENT CARS

The Columbia Six automobile, built in Detroit in 1916-1925, was an "assembled" car composed of parts from many reputable suppliers. The Columbia claimed to be "The Gem of the Highway" with such novelties as thermostatically controlled radiators, large 14-inch brake drums and "non-synchronizing spring suspension" which guarded passengers from "vitality-sapping road shocks." The 1922 model had a 50-horsepower Continental "Red Seal" engine, Durston transmission, Timken axles and bearings, Auto-lite electrical equipment, Borg & Beck clutch, Spicer universal joints, Gemmer steering gear, Harrison radiator and a sedan body built by Erdman-Guider. Advertisements stressed that Columbia was an "assembled" car built of "the best component parts money could buy, made by specialists in their fields."

As it turned out, however, buying all or most of the component parts from outside suppliers was economical only as long as the completed car could be sold at a reasonable profit. The cars of mixed ancestry did not do well when competition became tougher and price-cutting began. By the mid-1920s, when the postwar boom became less frenzied and people grew more particular about what they bought, many became suspicious of "assembled" makes, thinking they were not well designed or manufactured.

That fear was often groundless because the parts usually were well-known brands supplied as original equipment for many other makes of automobiles. The record shows that, while the number of motor vehicle producers declined steadily in the 1920s and succeeding decades, the parts and material suppliers serving Detroit have continued to number in the many thousands, even as the 21st century approaches.

Today, encouraged by both auto company downsizing and globalization, the supplier network is a mixture of old names including Budd, DuPont, General Electric and Solvay, relatively new names like ASC, Automotive Industries and Delphi, and a host of foreign-based companies such as Bosch of Germany, Lucas of England and Nippondenso of Japan.

Environmental regulations impacted the auto industry in both its products and its plants. Fifty-thousand-mile durability of vehicle emission controls required extensive use of computerized laboratory test cells such as this. (General Motors)

Going, Going, Green

Environmental policy imposes constraints, invites innovation, incites opposition

by J. R. Wargo

In 1995, the still relatively new engineering discipline of dealing with environmental concerns is one of the most complex, challenging and politically-charged in the profession. It involves reducing or removing real or perceived threats to the environment and to human health. It demands new assumptions and new approaches to product design, manufacture and application as well as plant design and operation. It involves today's engineers with monitoring, measuring and controlling plant and product effluents and choices on energy consumption and conservation that many of their predecessors didn't even know existed. This is done in a climate of changing federal and state regulations and standards, and against a background of unrelenting debate over whether these should be stiffened, left alone, or scrapped. Following is a brief history of some events, centered around automobiles, which prompted the development of this exacting discipline.

In 1895, and for decades to come, factory smokestacks belching black, sulphurous smoke were signs of growth and unstoppable progress — the American Way. Today, anything even remotely gray coming from a smokestack can prompt instant litigation from environmental groups. Conventional wisdom a century ago, and for some time to come, held the land as there for exploitation and the nation's natural resources as inexhaustible.

At the turn of the century, what was to become known as the environmental movement centered on three men: John Muir, Gifford Pinchot, and Theodore Roosevelt. The latter two worked as a team. Pinchot headed the Forest Service of the U.S. Department of Agriculture in the Roosevelt Administration. Between them, they assigned a top priority to protection of federal lands and water resources to assure their availability for future generations. Their goal — conservation, labeled at the time "the gospel of prudent use." Sierra Club founder John Muir preached what is now termed the biocentric view; all life has a right to exist, and to destroy plants or animals for personal gain is somehow immoral. His objective — preservation, or the protection of nature as is.

Conservation was more appealing to lawmakers, but only marginally. Often begrudgingly, legislation was passed soon after the turn of the century, and later enlarged upon, geometrically expanding what previously had been a token system of national parks and forests. Next came passage of an act to protect inland waterways. Having won on trees and streams, Roosevelt extended his goal. In 1908, he invited the governors of all the states to a White House Conference on Conservation. It resulted in a declaration that the protection of human health is an inherent goal in conservation. Two generations later, Muir's philosophy of preservation would come into its own as the environmental movement, and when it did it included but went beyond the 1908 Roosevelt expansion. It sought to conserve not only trees and fish, but all life, including human life. The movement seeks a system aimed at protecting this life from anything that might threaten it.

PROSPERITY FIRST, EARTH SECOND

In the next 60 years, through two world wars, a 10-year depression, and finally the balmy economic boom years of the Eisenhower era, the emphasis was on increasing production. Environmental concerns were not congenial to the thinking of the times, particularly if they threatened interference with productivity. In the 1960s, this mindset began to change.

April 22, 1970, when Earth Day observations were held in cities and at universities across the land, is considered a watershed date in the history of the environmental movement. That's the date that environmental awareness entered the mainstream national consciousness.

But what could be seen as a national indifference to ecological concerns actually began to melt in 1962, when marine biologist Rachel Carson's Silent Spring was published; first in condensed form in the New Yorker, and then as a thin book. In its own way it was as seminal as Harriet Beecher Stowe's much thicker

Uncle Tom's Cabin. Carson focused on a narrow range of chemical pollutants and their effect on the environment. It was her thesis, expressed in a spare style that bordered on the poetic, that the poisoning of the earth by chemical pesticides, most specifically DDT, threatened the destruction of nature and life. Her calm and reasoned exposition of available scientific data traced the flow of chemical contaminants in water and air through the food chain, eventually reaching men, women and children, shortening their lives and posing grave genetic danger to future generations. What kills flowers and trees, animals and birds, she reasoned, will in time kill human beings as well.

Carson was not the first to warn against DDT, but she was the first to do it in warm, moving prose, and she struck a chord that resonated rather than fading away. Following her book's publication, ecological insults were no longer regarded as isolated events, but rather as parts of a tragic pattern: man's sometimes deliberate sometimes inadvertent destruction of planet earth. A huge oil spill from an oil rig off the cost of Santa Barbara, Calif., the stagnant inversion of foul air over Donora, Pennsylvania, the discovery that toxic PCBs were being dumped into the Hudson River, the contamination of food fish by mercury discharged by industries into waterways, and the day that the Cuyahoga River in Cleveland caught fire because of the heavy concentrations of industrial chemicals ... these and other incidents were no longer treated as local phenomena but as mounting evidence that a pattern of abuse of the environment reflected a national policy of indifference. Slowly at first, environmentalism emerged as a political issue. Then it escalated.

ENVIRONMENTAL FRENZY

In 1966, Congress passed the National Traffic and Motor Vehicle Safety Act, and a year later, the first version of the now much-amended Clean Air Act. In 1969, it passed the National Environmental Policy Act requiring the federal government to analyze and report on the environmental implications of its activities. In the following year, Congress created the Occupational Safety and Health Administration, the National Institute of Occupational Safety, and the Environmental Protection Agency which set safety standards for workplaces. Within the next two years, Congress enacted the Federal Water Pollution Control Act, a sweeping pesticide control law, a noise control act, the Coastal Zone Management Act, and the Endangered Species Act.

Legislation passed by Congress in ensuing years stiffened the initial measures and added dozens, fulfilling a declaration by President Richard Nixon in his

February 1970 State of the Union address that the 1970s "absolutely must be the years when America pays its debt to the past by reclaiming the purity of its air, its waters and our living environment."

Whether it reflected enlightenment or political savvy, this statement was made weeks before Earth Day and the nominal launching of the environmental movement. In a sense, then, victory was in hand before the war began. Ironically, the movement was merely in a larval stage by the time much of the basic environmental legislation had been enacted into law. The first of the new-style environmental groups, the Environmental Defense Fund, wasn't founded until 1965. After a group of Long Island, N.Y. lawyers sued successfully to block local use of DDT, they decided to form a "club." From the start it was a different type of group: it introduced enduring innovations in the use of science and the law to support demands for ever stiffer regulations and for enforcement of those already on the books.

Soon there were hundreds of environmental groups — now thousands if one counts local grass roots groups — some formed around a single issue, others lobbying, litigating and educating the general public on a broad range of environmental issues. Where did they all come from? Some, such as the Union of Concerned Scientists, began as an antiwar group, fighting to halt the conflict in Vietnam. When the war began to wind down in the early 1970s, its minuscule staff was faced with either closing up shop or finding a new issue. They opted for the latter, selecting commercial nuclear energy as their second target, generating intense public scrutiny of the industry. The UCS staff of dozens now concentrates on a dozen projects, from insecticide-free agriculture and transportation to international arms control and renewable energy.

GRASSROOTS GROUPS

Other groups were spawned by organizers who, during the Vietnam debate, became convinced that special interests, particularly big business, controlled the nation's government. To counter this influence, they formed citizen action groups to protect the interest of the individual. In the minds of many activists the environmental movement is a device for introducing true democracy to a republic that has evolved into an oligarchy.

Old-style conservation groups — the Sierra Club, the Wilderness Society, and the National Audubon Society, to name but three — suddenly found themselves anachronisms. They reorganized, adopted the tactics of the civil rights and antiwar movements being used so effectively by the new organizations, and made the transition into the environmental movement. Thus redefined, they have grown to a level of prominence and have attained a degree of political clout their founders could never have imagined.

One view of the sociological and political phenomenon classes the environmental movement as a series of pitched battles over technological issues waged by idealists who understand their objectives only vaguely and have an even flimsier grasp of what is involved in achieving those objectives. This view is often put forth by those with technical expertise who, for the last quarter century, have been tasked with revamping products and production methods to meet the flood of federal regulations that the movement helps spawn. One of the earliest challenges for these experts was emissions control.

In 1970, the Clean Air Act was amended to require a 90 percent reduction in automotive emissions, to be reached incrementally. At the time the amendment was passed, the technology to meet the first increment — due in 1975 — did not exist for cars produced in high volume. Eventually it was revealed that the 1975 date for the standards was selected by the congressional staff preparing the amendments by the simple process of subtracting 5 from 1980, the date for those standards in an early draft. When auto industry representatives petitioned for review and revision, the U.S. Environmental Protection Agency came up with interim standards by the simple process of subtracting what the industry was requesting from the original 1975 standards, splitting the difference and tacking it on to the industry request.

The state of technology was not considered in the technological demands being inflicted by law on industry. Automotive engineers were faced with politically imposed deadlines for technological breakthroughs. This is not the way, some argued, that science and engineering usually works.

DRAWING A LINE

In a loose sense, not a few environmental regulations were imposed on industry in the same arbitrary manner. In defense of the process it is argued that industrial contamination of land, air and water, and product designs that often threatened human health, were obvious, and the process of correcting this abuse of land and life had to begin somewhere. Standards in the form of mandatory numbers were the most logical starting point to get industries moving. Self-regulation obviously wasn't hacking it.

Air pollution from cars is a good example. After the first major California smog experience in 1943, the state assembled academicians, lawmakers and auto engineers to examine the problem. The principal blame fell on the car. The study sat on the shelf. The first federal research program on air pollution began in 1955, and the first national conference on air pollution sponsored by the Public Health Service was held in 1958. The car was singled out in each instance as the major contributor. But the auto industry didn't react until 1960, after California set auto emission standards requiring crankcase blow-by valves on all engines.

Even then, the auto industry's first reaction was to resist. In 1966, the National Academy of Sciences reported that transportation contributed 59.9 percent of the 125 million tons of principal pollutants dumped into the air annually, and cited the automobile as the chief offender. So air quality standards were set — albeit somewhat arbitrarily at first — however, not only for the automotive industry but for all types of manufacturers as well as coal- and oil-burning utilities.

In hindsight it can be stated that automotive industry resistance compounded the burden on engineers tasked with meeting new air and water quality and product safety standards. Time and money which could have gone into advancing existing technologies was diverted instead to developing a defense of a basic industry posture that the standards were either unachievable or achievable only at a cost destructive to the economy. And there was another basic tactical error.

To some observers, carmakers appeared to have entered the age of environmentalism and consumerism so convinced of the nation's love for their product that they were slow to realize the seriousness of the crisis they faced. Their Washington lobbying offices were soon to be skeletal when compared to the staffs some activist groups were building up to influence Capitol Hill. Long into the game, carmakers still regarded car owners as their best lobbyists.

Consciousness Raising

Activists, however — to some extent armed with lessons learned while teaching the general public to loathe U.S. involvement in Vietnam — created a grassroots movement that soon grew into a groundswell of public opinion in support of demands that all industry protect the environment and public health, and that the government force industry to so act. Americans may love their cars, but activist citizen education programs created a consensus that they love clean air and safe products more. Elected officials were quick to sense where the public was coming down, and equally quick to give them what they wanted.

But the influence activists had on the national attitude towards cars was transcended by the impact of the two oil embargoes during the 1970s. Many motorists reacted by becoming more resentful of the makers of cars, whose products seemed to visit gas stations too often, than they were of the Arab states which had temporarily turned off the tap. One automotive writer at the time described his job as "covering the auto industry's painful transition from enthronement as a mass love object to oblivion as a costly anachronism." Radical activists responded to the embargoes by demanding a phase-out of gas burning cars. Moderate groups lobbied Congress for mandatory curbs on gas consumption and compulsory marketing of natural gas-powered and electric vehicles. Congress, however, was already working on the former, and California was eventually to come up with the latter.

The congressional solution, supported by all mainstream environmental groups, was passage in 1975 of legislation mandating establishment of corporate average fuel economy (CAFE) standards, currently 27.5 miles per gallon. Environmentalists gave full support to legislation introduced in 1990 by Sen. Richard Bryan of Nevada that would have boosted the CAFE standard to 40 mpg by 2001. Industry representatives like to point out that Bryan drives an Oldsmobile averaging less than 20 miles per gallon, proving only that the quality of the cars made in the last twenty five years has improved considerably more than the quality of debate over federal actions mandating that improvement.

In California, which led the way in 1960 with the first air pollution standard, the state Air Resources Board in 1990 approved rules requiring that, by 1998, 2 percent of all cars sold must be ZEVs (zero emission vehicles), rising to 10 percent in 2003. The standard, in effect, mandates electric vehicles. Massachusetts and New York have also passed laws promoting the sale of ZEVs. Activist groups are asking members in other states to lobby for the passage of similar legislation, making it the issue of the moment.

Almost from the start, a quarter of a century ago, environmental groups established a symbiotic relationship with liberal elements in government. The groups would generate ideas, and the legislators would pass legislation incorporating some of these ideas, although but a small fraction of those proposed. The more significant impact of activists has not been in influencing legislation but in forcing the implementation of the legislation, either through litigation or working with

ELECTROMECHANICAL CARBURETOR

ELECTRONIC CONTROL MODULE (ECM)

THROTTLE POSITION SENSOR

EST DISTRIBUTOR

DIAGNOSTIC LIGHT

IDLE SPEED ACTUATOR

MANIFOLD PRESSURE SENSOR

DUAL BED CATALYTIC CONVERTER

CHARCOAL CANISTER PURGE

COOLANT SENSOR

AIR PUMP & MANAGEMENT VALVE

OXYGEN SENSOR

DUAL BED CATALYTIC CONVERTER

3-Way Catalyst   **Air Chamber (between beds)**

ENGINE EXHAUST

INDUCTED AIR

By the early 1980s, control of vehicle emissions to comply with increasingly stringent federal and California environmental regulations required an array of components on passenger cars, as exemplified by one company's 1981 system. (General Motors)

federal regulators to develop standards which fulfilled the mandate of the legislation. No less significant is the network of thousands of exisiting grass roots groups. Its members can be counted on to churn out tens of thousands of letters, calls, faxes and E-mail messages to members of Congress when a vote judged critical to protection of the environment goes to the floor.

REPEAL AT OWN RISK?

The installation of a conservative Senate and House at the start of 1995 raises in some the hope, and in others the fear, that much environmental legislation and implementing regulation will be rolled back. The general public, however, is by and large fairly comfortable with what has been achieved in the name of environmentalism, with significant achievements in cleanup and protection able to be cited. While a little tinkering with environmental protection measures may be tolerated, undoing it to any significant degree likely could be the undoing of those legislators responsible come the next election. (A 1989 poll showed 80 percent of voters regard themselves as environmentalists.)

An awful lot of environmental rhetoric has been silly hype. The theater of the more radical groups has often been embarrassing to watch. But the basic objectives of the moderate, mainstream groups — federal action to preserve the environment, and expanding the voice of the individual citizens in determining how natural resources are to be used and how their personal health is to be protected — were long overdue, have been achieved and, like it or not, are gradually becoming integrated into the political and social fabric. The citizen involvement aspect has generated a backlash from the right, so far mostly over land use issues. A lot more people on both sides who previously only sat on the sidelines now engage in healthy debate over issues affecting their lives, and influence policy in the bargain.

Despite increasing questions about the goals, methods and sometimes even the motives of environ-

mental activists, their organizations could evaporate in the next instant, and still there would be no going back to the way things were. In the wake of the environmental movement have come the technological advances which have done the actual job of making life better. No one is taking those away.

In The Blink Of An Eye
The Future Is Here

It was just a few years ago, 1938 to be exact, that Kolene founder, John H. Shoemaker, first commercialized the molten salt bath cleaning process. In 57 short years, we have watched our technology evolve from the laboratory process Mr. Shoemaker demonstrates above, to sophisticated processes for cleaning, conditioning and modifying metal surfaces.

Like ESD, The Engineering Society, which celebrates its 100th anniversary this year, Kolene serves the engineering and manufacturing communities by continuously challenging the limits of its technology.

Today, Kolene offers manually operated units, automated batch systems, and in-line configurations for integration with new or existing plant conveyor systems.

The application of Kolene salt bath technology has changed, too. Today, Kolene processes
- remove burned-in sand and surface graphite from cast iron
- descale stainless steels, superalloys and refractory metals
- remove shells and cores from investment castings
- strip paint buildup from hooks, racks, fixtures and parts
- improve the wear, fatigue and corrosion resistance of ferrous components.

In the blink of an eye, the future of Kolene salt bath technology is here. To learn more about our capabilities, contact Kolene Corporation, 12890 Westwood Ave., Detroit, MI 48223. Call us at (313) 273-9220. Toll free: **1-800-521-4182.**

KOLENE® **CORPORATION**

III. ₀ 'Systems Analysis'

Technological advance, demographic change, rise of infrastructure intertwine

WHAT IT COSTS.

HOW IT PERFORMS.

When it comes to making material choices in the 90s, cost is a big consideration. But a low-cost choice is no bargain if it compromises the quality of the molded fascia. ■ Now there's a high-value answer. New RIM-Lite from Bayer Corporation. It's everything you want in a bumper material. And it costs about 15% less. ■ Let's talk performance. New RIM-Lite gives you all the advantages of regular RIM polyurethanes. Like easy paintability and superior design flexibility.

NEW RIM-LITE BUMPER MATERIAL FROM BAYER. POLYURETHANE PERFORMANCE AT A FRACTION OF THE COST.

■ Wait, it gets better. Because of a breakthrough in polyurethane chemistry, RIM-Lite actually improves a bumper's impact resistance. And our new low-density system gives you a finished part that weighs about 15% less. Making RIM-Lite cost-competitive with some other materials that, quite frankly, aren't in polyurethane's league. ■ Want to hear more? Contact Bayer Corporation, Polymers Division, 100 Bayer Road, Pittsburgh, PA 15205-9741. 1-800-662-2927.

Bayer

Polymers Division

On April 3, Miles Inc. became Bayer Corporation.

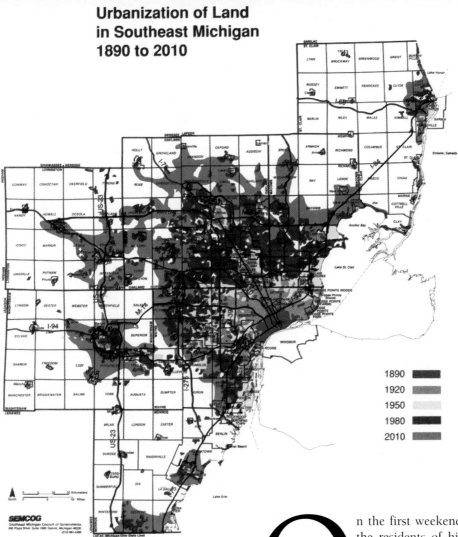

Urbanization of Land in Southeast Michigan 1890 to 2010

1890	■
1920	■
1950	■
1980	■
2010	■

SEMCOG
Southeast Michigan Council of Governments
660 Plaza Drive Suite 1900 Detroit, Michigan 48226
(313) 961-4266

Boom & Bust

*Census, cycles of industry
lie in complex — and sometimes
precarious — relationship*

by Maureen McDonald

TOP: *This map shows the patterns and periods of population growth in Southeastern Michigan over the last century. (Southeast Michigan Council of Governments)*

On the first weekend of June in 1995 the residents of historic Indian Village near Detroit's riverfront celebrated the 100th anniversary of their neighborhood. It was here in these spacious quarters that early automobile pioneers such as Horace Dodge and Edsel Ford took up residence, along with wealthy barons of steam engine, rail car and pharmaceutical manufacturing.

Over the years some homes became churches, boarding houses and community buildings. Active historic preservationists work to keep a chunk of the city relatively unscathed by the relentless winds of change. Few original neighborhoods have survived. Auto manufacturing and suppliers have transformed the city from the hub of auto manufacturing to the mecca of research and technology.

Its residents — empowered by the wheels they design and sell here — have pushed beyond the borders of Livingston, Genesee and St. Clair counties, growing subdivisions where corn and soybean fields once flourished. Better roads and quieter vehicles, equipped with cellular phones and CD players, have allowed people to commute up to two hours daily to work.

From humble dwellings clustered in a 12-square-mile semicircle bounded by Grand Boulevard and the Detroit River in 1890, the population in the 1990s fans

IF COLLISIONS WERE FUN, YOU WOULDN'T NEED ENERGY-ABSORBING FOAMS FROM BAYER.

Ah, the good old days. There you were, tucked behind the wheel of trusty number nine at the Bump-O-Car ride, eager to punch the accelerator. And race headlong into anything that moved. Or better yet, someone who couldn't. ■ *If only real crashes in real cars were enjoyable. Then you wouldn't need Energy-Absorbing Polyurethane Foam Systems from Bayer Corporation. But they're not. So you do.* ■ *Bayer engineers have simulated countless crashes in our lab. From the side. The front. The rear. And the resulting knowledge has helped us develop polyurethane foams that*

absorb energy better. To help make today's cars safer. ■ *When used in side impact bolsters, our EA Foams can help protect a car's occupants in side collisions up to 35 m.p.h. In knee bolsters and bumpers, they can help reduce the risk of serious injury. And when it comes to parts consolidation, EA Foams are an excellent choice.* ■ *Bayer EA Foams. While they can't make collisions fun, they can help make cars safer for everyone.* ■ *For more information, contact Bayer Corporation, Polymers Division, 100 Bayer Road, Pittsburgh, PA 15205-9741. 1-800-662-2927.*

Bayer

Polymers Division

On April 3, Miles Inc. became Bayer Corporation.

out to seven counties. Manufacturing followed, and sometimes led the charge. Job growth in suburban Oakland and Macomb counties increased 400 to 500 percent between 1953 and 1991, according to the Southeast Michigan Council of Governments (SEMCOG). Livingston experienced a sixfold job increase between the same period and Washtenaw tripled its employment. The city of Detroit languished as auto manufacturers and suppliers built newer plants in the suburbs and in other parts of the country.

Today sport enthusiasts play Splatball, a joyful form of capture the flag, in cavernous rooms of the abandoned Packard Motor Car Company. Rusting assembly parts are stored at the historic Ford factory in Highland Park. Yet new life can be found for old buildings. General Motors invested millions to transform its 4-million-square-foot Pontiac Central Truck Plant, a 1927 building, into the GM Truck Product Center, a place where 4,200 engineers would work. The push towards new, modern dwellings and work sites has been almost as rapid as automotive model changeovers.

Going Mobile

Sociologists cite factors such as a highly mobile population and a better-than-average middle-class wage provided by automakers and big suppliers in analyzing Detroit's unique demographics. Research shows that more metro Detroiters own single family homes than residents of other U.S. cities. Henry Ford himself invested in workers with marriages and mortgages, believing they tended to be more loyal to an employer and adherent to the work ethic than single apartment dwellers.

The city of Detroit reached a peak population of 1,910,000 in 1950, but that plummeted to 1 million by 1990. Between 1985 and 1990 more than three times as many whites moved out as moved in, according to SEMCOG. Meanwhile the regional population swelled to 4,660,897 in 1993.

What endures as metro Detroit's most recognizable trait throughout the decades that it is a city of capable workers — visionaries, modelers, workers, suppliers, marketers and repair technicians. Autos fuel the nation's economy, with at least 190 million motor vehicles registered in the United States. In 1994, the 22,850 U.S. dealers sold over 15.5 million new cars and trucks according to the American Automobile Manufacturers Association (AAMA).

Over 1 million Americans work in motor vehicle and equipment manufacturing, which along with related industries contributed a whopping $2.1 trillion to the annual economy in 1990, according to the U.S. Department of Commerce, Bureau of the Census. Metro Detroit continues to reap the benefits.

David Cole, director of the Office for the Study of Automotive Transportation at the University of Michigan, says southeast Michigan is rebounding as the epicenter of the automobile brain trust approaching the millennium.

"Now that the Big Three automakers have reformatted themselves and become highly agile and dynamic, their home town has become the dominant center for automotive knowledge and technology. Transplant automakers have put their research facilities here. International suppliers are flocking here to be close to the Big Three's technology headquarters. No other city in the world has this much magnetism and intellectual strength," says Cole.

Detroit was destined to become the town that put the nation on wheels, owing to its relation geographically to Canada and the Eastern United States and its versatile manufacturing economy, which was already booming at the turn into the 20th century, according to Robert Conant, author of American Odyssey. He noted that Detroit obtained an electric lighting franchise before New York City, thereby improving its manufacturing prowess.

Then came Henry Ford's Model T production. and the world-famous $5-a-day wage.

Historian Melvin G. Holli says that on the eve of the manufacturing "takeoff" in 1904, the foreign-born constituted about one-third of Detroit's population, among whom were 13,000 Poles, 1,300 Russians and 904 Italians. The 1910 foreign-born population of 156,365 doubled to 289,297 in 1920. By 1925 the foreign born constituted about one half of Detroit's 1,242,044 people, the Poles leading the list with 115,069. Italians then numbered 42,457, Russians 49,427 and from almost a zero baseline Hungarians had come to number 21,656, according to Holli in his book Detroit, the Documentary History of an American City.

Early Urban Flight

The population explosion was unparalleled anywhere in America. Holli notes that Detroit jumped from 13th to 4th place in population among major American cities between 1900 and 1920. The impact of migration disrupted Detroit's social patterns, stretched its boundaries, crowded its housing, created new sanitary, health and education crises, but also brought fabulous wealth to automotive and real estate investors and the city itself. Wealthy executives began moving to Grosse Pointe in the early 1920s while immigrants made rooming houses out of many downtown mansions.

The city of Detroit was referred to as "Utopia on Wheels" until the 1930s. Of the 2,579 "substantial families" registered in Dau's Blue Book in 1910, 52 percent lived within three miles of the downtown business section. By 1930, fewer than 8 percent lived near downtown and 50 percent lived beyond the city limits.

The push to Bloomfield Hills began in earnest in the 1950s.

"The boom and bust cycle tended to be exaggerated in a one industry town," explains Holli. The number of Detroit wage earners in manufacturing fell almost 40 percent by 1933. Automobile workers led the downward spiral with nearly one-half of them idle during the Depression.

In 1932 the "Ford Hunger March" garnered worldwide attention. "America had come asking for help in frayed trousers, cracked shoes and belts tightened over hollowed bellies," observed novelist John Dos Passos, "and all they could think of at Ford's was machine guns."

Of 3,000 hungry — and Communist-inspired — marchers who demonstrated outside the Rouge Plant, four were killed by the Dearborn Police in a large melee. But the resolve to form labor unions was strengthened. By 1936 Detroit's plants were beset with walkouts, wildcats and sitdowns, most notably at GM's Flint Assembly Plant. The United Automobile Workers became the recognized bargaining agent for GM in 1937 and the industry at large over the next four years.

A catalyst for worker-company cooperation was the looming war effort. As industrial Detroit became the touted "arsenal of democracy" for World War II, companies acquiesced to some union demands and reached out to a broader circle of hirees, including 75,000 black workers, primarily from southern states. Negroes, as they were known in those days, had gained a foothold in Ford's foundries in the 1920s, to which they had been recruited by several local clergy. Now they were truly needed.

BATTLE OF BELLE ISLE

White opposition was fierce, according to labor historian Steve Babson in Working Detroit. Hate strikes, sitdowns and walkouts — mainly at supplier plants — culminated in a bloody riot on June 20, 1943. Up to 5,000 people were embroiled in a battle beginning on Belle Isle. At its end, 25 blacks and nine whites were dead and 700 people of both races were injured. On the positive side, blacks had become a formidable part of the labor force, creating the wealthiest black middle class in the nation. By 1995 Detroit also had the largest number of elected black leaders in the nation.

The racial tensions eased but the pace of white out-migration increased measurably after World War II, fueled in part by Veteran's Administration mortgages and the lure of new satellite shopping malls, the first located in Harper Woods (Eastland) and Southfield (Northland).

Robert Sinclair, a Wayne State University urban geographer, observes that Detroit's ethnic groups — as in other cities — moved along set migratory pathways radially into the suburbs. Hungarians moved out from Delray into downriver communities such as Southgate and Wyandotte along the Jefferson corridor. The Polish population moved beyond Hamtramck along Van Dyke in an eastern corridor, and the Irish from Corktown to the University of Detroit neighborhood and thence scattered.

The Jewish population moved north along the Northwestern corridor settling first in the Dexter-Davison area of northwest Detroit, then in Oak Park and Southfield and eventually Franklin, West Bloomfield and Farmington Hills. The German population started in Harmonie Park and moved west along Michigan Avenue to Dearborn Heights and Livonia and eastward out Gratiot. The black population started in Paradise Valley, a confined area on the near east side destroyed by I-375, and sprawled to the far corners of Detroit, Oak Park, Southfield, Inkster and River Rouge.

The mid-'50s were also a high-water mark in terms of automotive employment, according to Holli. Big sales years such as 1955 and 1965 escalated wages and incomes, filled normally vacant housing, and attracted to the city new workers who were often unemployable under more normal conditions. Down-years such as 1958, 1974 and 1979 represented near-depression in Detroit during what was only recession elsewhere in the nation. The major decline in Detroit employment reflected reduced defense spending, an increase in efficiency and substitution of imported car and truck sales.

The combined impact of shrinking employment and social unrest led to brutal racial tensions. On July 23, 1967, the city was once again on fire. Over the next four days it took more than 10,000 state and federal troops to quell the fighting. By the time the smoke cleared, 43 people were dead, another 347 were seriously injured and over 7,000 were under arrest. Some 1,300 buildings lay in ruins. Compounding the sense of fear among white residents, a federal judge ordered Detroit Public Schools desegregated in 1972, which led to citywide busing.

The sense of danger and frustration was ominous. In 1973 the number of homicide victims in Detroit was triple the death toll on all sides in the civil disturbances that took place in Northern Ireland the same year. Thousands of guns were sold to Detroiters of every race and class who sought protection from social chaos, according to Dan Georgakas, author of Detroit Do Mind Dying.

Meanwhile, the racial tensions inside the automotive plants were beginning to boil. On July 15, 1970, a black auto worker named James Johnson entered the Eldon Avenue Gear and Axle plant of Chrysler Corporation with an M-1 carbine hidden in the pant leg of his overalls. He wound up shooting one black foreman, one white foreman and one white job setter, one of the first cases in an incipient trend toward violence in the workplace. Within the next 25 years at least eight inner city manufacturing plants would close and Chrysler would move its headquarters to Auburn Hills. By 1995 Chrysler and GM each had only one assembly plant in the city, subsidized by city and federal grants, and Ford had none.

Rust Belt Resurgence

The recession of the 1980s did more than devastate the Motor City, it caused government and employment leaders to re-think its future. Increasing competition from Germany, Japan and other nations spelled a loss of luster as well as production for Detroit. Some 115,000 manufacturing jobs have been lost since 1970, according to SEMCOG. Such jobs constituted 40 percent of the region's employment in 1965, dropped to 24 percent in 1990 and were projected to drop another 17 percent by 2010. Still, the 1990s saw the beginning of a resurgence in research, technology and training jobs related to the automobile industry.

Regardless of turmoil, boom and bust periods, or lure of warmer climates, Detroiters still wax optimistic about their chief industry, which has endured a full century. The freeways are named after car makers and union leaders. The Detroit Historical Museum and Dearborn's Henry Ford Museum pride themselves on their automotive exhibits. The rejuvenated North American International Auto Show in Detroit ranks among the best attended in the world.

The sense of hard work and craftsmanship remains endemic to metro-Detroit personalities. Call it "gasoline in the blood." Illustrating this are three views of the nitty gritty city and its chief employer:

The zest for craft was spoken most eloquently by a fictionalized car designer in Arthur Hailey's book Wheels:

"You have to love cars to work here. You have to care about them so much that they're the most important thing there is. You breathe, eat, sleep cars, sometimes remember them when you're making love. You wake up in the night, it's cars you think about — those you're designing, others you'd like to. It's like a religion. If you don't feel that way, you don't belong here."

But the anguish over speedup work detail and the relentless pace of the line was most clearly spoken by a worker at GM's Lordstown plant to an interviewer in the New York Times, Sept. 9, 1973: "They tell you, 'Put in 10 screws,' and you do it. Then a couple of weeks later they say, 'Put in 15 screws,' and next they say, 'Well, we don't need you no more; give it to the next man.'"

One of Detroit's most famous novelists, Elmore Leonard, sums it up this way, in his introduction to the photography book Detroit: The Renaissance City.

"There are cities that get by on their good looks, offer climate and scenery, views of mountains or oceans, rockbound or with palm trees. And there are cities like Detroit that have to work for a living."

By Design or Destiny

*Detroit architects, builders left mark on the landscape
— and their profession*

by Jim Gallagher

At the end of the 19th century, three major forces were converging on American architecture and construction, and nowhere were those forces more influential than in Detroit. The first was the Classical design influence of the 1893 Chicago Columbian Exposition into the design of all forms of buildings, commercial and residential, at every scale of size.

The second was the introduction of structural, mechanical and electrical engineering advances as integral elements of new buildings. The third was perhaps most important here in Detroit: the birth and

Technology-Driven, Customer-Focused

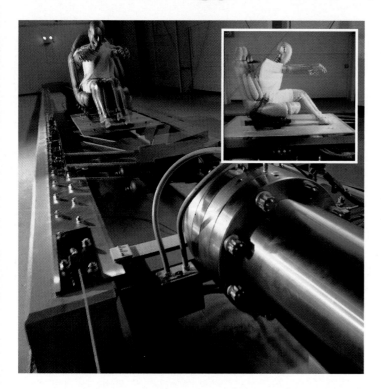

Just as Lear Seating Corporation is the world's largest independent supplier of just-in-time automotive seat systems, its Engineering Technical Center is an industry leader in design and manufacturing technologies. High-quality, low-cost seating is the common customer-focused goal, driven by these components:

Test and Development Laboratory
• HYGE crash simulator • Multi-axis shaker table (MAST) • On-road data acquisition
• Environmental chamber • Fabric testing
• Lab and in-vehicle modal analysis
• Hydraulic strength testing • Pneumatic fatigue testing • Lab view computer systems
• Acoustic data acquisition • Zwick indention load deflection • Jounce and squirm • Sliding entry.

Seating Concepts
Benchmarking Laboratory translates industry-best measures and practices into seat design guidelines. *Biomechanics/Ergonomics Laboratory* uses the sciences of Body Pressure Distribution, Market Research and Anthropometry to study the correlation between subjective and objective comfort evaluations, based on a current database of 3,000 seat-subject combinations.

Product Development
Advanced Engineering utilizes new technologies and applications to generate innovative, customer-focused products and services. *Product Design* transfers those new technologies throughout Lear's global network.

Foam and Trim Development
Foam Laboratory produces prototype seat foam using production-simulation processes, and validates new model tooling. Ongoing development of chemical formulations and materials is integral to prototyping. *Trim Laboratory* uses advanced technology — such as pattern digitizing and nesting, Pixsys 3D-CAD scanning — to develop and prototype patterns to meet design intent. *SureBond™ Laboratory* evaluates fabric feasibility and establishes process parameters for this patented fabric-to-foam bonding system.

Technology-driven, Customer-focused at 79 locations in 17 countries.

Seating Corporation

World Headquarters
21557 Telegraph Road
Southfield, MI 48034
810/746-1500
Member NYSE

growth of the automobile industry with its demands for enormous clear-span buildings, capable of enormous loading, and requiring unlimited potential for changes in use.

Detroit was particularly well equipped to meet these new challenges. The leading local designers were either European- or Eastern-trained, or had long hard training under older men who were so trained. Yet, they looked West for inspiration for the new, not back to the East for the past. They had extensive architectural libraries to call upon.

The larger firms were architectural/engineering in structure, following the Chicago pattern, rather than the architecture-only organization of New York and Boston. Three of the four founders of Smith, Hinchman & Grylls were recent graduates of the University of Michigan's engineering college. Another giant-to-be, Albert Kahn, was shaped by his structural engineer brother Julius, who brought his pioneering knowledge of reinforced-concrete buildings, perfect for the new factories.

Within a few years after the landmark Packard Motor Car Co. reinforced-concrete buildings in Detroit (1905), and the steel-framed Olds Motor Works in Lansing (1903), huge new factories for the production of autos and their parts were appearing throughout southeast Michigan and the entire Midwest, designed by firms who had all of the design and engineering in-house.

TRIPLE LEG-ACY

But design is but one of the three legs of any building, construction and labor being the others. Two of Detroit's largest 19th century industries had been shipbuilding and railway car manufacturing. Although both these industries were on the down-side of their boom years, the men who built the shipyards and the railroad-equipment factories were neither afraid of the large scale of the auto facilities needed nor were they lacking in the skills to build them. Skilled workers from every country in Europe had been pouring into the United States by the millions, eager for the opportunities that the burgeoning nation promised them. Anything you could design and blueprint they could build.

PAGE 95: *Design of GM's Technical Center in Warren, shown here at its May 16, 1956, dedication, was a major accomplishment of the decade for Smith, Hinchman & Grylls (SH&G) and architect Eero Saarinen. (General Motors)* BELOW: *This view of the Detroit of the 1990s illustrates the lasting influence of SH&G's designs on the core area. In the foreground are, l. to r., SH&G's Penobscot Building (1928), Buhl Building (1925), Guardian Building (1929) and the lace-like Michigan Consolidated Gas Building (Minoru Yamasaki, 1961). (SH&G/Balthazar Korab)*

Delivering Competitive Technology

To compete in a global market, automakers constantly search for new manufacturing ideas, pick the most practical, and implement these ideas, fast. *And DCT provides the innovation to help.*

DCT is a global leader in automotive manufacturing innovation. Our global knowledge and reach allow us to provide systems and services wherever our customers manufacture their product.

DCT provides state-of-the-art technology to automakers who benefit from our ability to simplify turnkey manufacturing systems with advanced technology and design methods. Advanced systems made simple and practical with hands-on experience and superior technology. That's innovation.

Innovation, before you ask! ®

DIVISIONS INCLUDE:

DCT Welding & Assembly
 automated welding systems engineering
DCT Material Handling
 material handling solutions
DCT Component Systems
 die design and production stampings
DCT Fasteners
 innovative fastening technologies
DCT Advanced Engineering
 research and development
Utilase
 laser systems
 tailored blank welding
Stature Prototyping Services
 3-D parts from CAD data
Control Power-Reliance
 testing systems

By 1916, the major contractors in Detroit were strong enough to form a trade organization, Associated General Contractors of Detroit, which within two years had joined a national group of similar contractor associations with the same goals. This Detroit group has continued for more than 75 years, and lists as members firms that have carried out major projects, locally, nationally and throughout the world.

During World War I, for the Allies and later the U.S., the still new automobile plants were greatly enlarged and began turning out the huge quantities of armaments needed. Schedules were greatly accelerated, the labor force grew to meet the need (and the city exploded with migrants from all over). The enormous Highland Park plant of Henry Ford became obsolescent almost overnight and began to be replaced by the even more gargantuan Rouge Plant in 1917.

This behemoth was soon able to provide every component that went into an automobile, including the steel and the glass. The saying was: "Iron ore and sand in one end and cars out the other."

With the end of the war and the mushrooming of demand for autos and other consumer goods, huge complexes like Ford, Dodge, Main and Packard were joined by the only slightly smaller Hudson, Chrysler, Continental, Briggs, Kelvinator and a host of others. Another industry, drug manufacturing, led by Parke-Davis and Frederick Stearns, built multi-storied facilities near the Detroit River.

Building Boom

The 1920s, for Detroit as for the entire country, were boom times, and the downtown streets were soon shadowed by towering skyscrapers. Housing spread to the outer city limits and streetcar lines crisscrossed the city, providing the link between residence and employment for all. At the peak of the boom, Detroit was third only to New York and Chicago in construction volume.

The Buhl Building, the Penobscot, and the Union Trust (now the Guardian Building) all went up in a space of four years, all by Smith, Hinchman & Grylls, and all on Griswold Street. Albert Kahn weighed in with several towering buildings, capped by the Fisher Building in the New Center area, two miles to the north, which joined his massive General Motors Building in the city's "second downtown." Smaller firms, headed by talented designers, were responsible for such handsome office towers as the Book Building and the David Scott Tower. These same firms also designed handsome Revival-style residences, houses of worship and country clubs for Detroit's elite.

The "golden '20s" were just that for architecture and construction in Detroit, but the gold turned to dross in 1929 with the onset of the Great Depression. Projects under way were halted and new plans were abandoned. Major contractors and experienced architects were soon without commissions of any size or kind. The big mansions, the factories, the office buildings, and the elaborate churches were no more, and builders and architects tried to stay alive with a tiny rag-tag of remodelings and maintenance work. The largest firm in the city, in 1932, booked only two projects, a doctor's clinic and a log-cabin vacation home.

The Depression almost wiped out the architectural and building industries in Detroit. Firms with generations of success and experience shrank to minuscule size or vanished completely. Albert Kahn sent a team of his best architects, engineers and superintendents to Russia for several years, assisting in the construction of more than 500 factories in the U.S.S.R., which turned out to be vital in the war that loomed ahead. Smith Hinchman & Grylls benefited from the repeal of Prohibition and the expansion of Hiram Walker's distilleries in the U.S.

Arms and The Men

The onset of World War II brought an industrial (read "arms") reconstruction that, beginning about 1938, led to the biggest change in Detroit since the rise of the automobile. This time the growth was on the western and northern edges of the city, since there was no longer the vacant acreage within the city that these armament facilities required.

One after another, these giant new factories turned Detroit into the "Arsenal of Democracy." Tanks, guns, planes, trucks, and shells poured out of both old and new buildings. What had been thought of as the city of Detroit became the tri-county area of Wayne, Oakland and Macomb Counties for construction purposes. The almost universal ownership of automobiles made it possible for employes to reach these outlying facilities. They were no longer tied to streetcar lines, and highways replaced rails as the link

The end of the war saw many of these plants reconverted to the production of autos and other consumer goods. The continued move of the city's population center to the outskirts now included the retail shops (which gleefully moved to the new shopping centers), endless subdivisions of single family homes, and finally, office buildings. Partly this reflected the suburbanization of family life (fueled by the returning servicemen and their new families), and partly the result of

The Great Depression of the 1930s was especially hard on architects and big-building builders. One of the few bright spots was this Wyoming Road plant for the DeSoto Division of Chrysler Corporation (1936) by Albert Kahn. (Albert Kahn Associates)

new highways that made access to distant points commonplace.

The chief victim of this change was the central city, which, within two decades, lost all three of its department stores, many of its office and clerical jobs, and square miles of its older neighborhoods. This was partly offset by new governmental buildings: a convention center and arena, a veterans' building, a new auditorium, and a mammoth city-county building, all on prime water frontage. But behind this facade — almost nothing.

The streetcar and bus lines that once connected neighborhoods with downtown shrank, then disappeared. Construction continued at a rapid pace, but instead of being concentrated around lower Woodward and the New Center area, it gave birth to even larger "downtowns" in the suburbs of Troy, Southfield, Dearborn and Bloomfield Hills. The argument can be made either that jobs followed the population or population followed the jobs, but either way, the relocation was followed by roads and infrastructure, shops, schools and churches.

Ahead of the Curve

About this time, two noted architects from the area, Eero Saarinen and Minoru Yamasaki, extended their reputations and their practices nationally, and then internationally. Saarinen, with his father, Eliel, won acclaim for their institutional buildings at Cranbrook, and then the $100 million General Motors Technical Center, on 900 acres of farmland northeast of Detroit. This project brought Saarinen acclaim throughout the world and commissions from some of the largest U.S. corporations. His career culminated in the TWA terminal in New York and the Dulles Airport near Washington, as well as the breathtaking Jefferson Memorial arch at St. Louis, a 590-feet high stainless steel cautionary curve.

Minoru Yamasaki was brought to Detroit as head of design for Smith, Hinchman & Grylls, but soon found a large firm too restrictive. With two coworkers, George Hellmuth and Joseph Leinweber, he moved to St. Louis and designed the barrel-vaulted St. Louis Airport Terminal, an instant success aesthetically and functionally. He moved back to Detroit, established his own firm, and commissions poured into the office from around the country and the world.

Although the larger firms in Detroit had designed major projects in many cities for their industrial clients, it took these two "form-givers" to really call Detroit to the attention of the architectural world. These large firms still call Detroit home, and are in constant demand for some of the nation's largest projects, partly because they offer "one-stop shopping" to their clients with all of the engineering and planning disciplines in-house.

One exception to the suburban migration of important buildings was the gigantic Renaissance Center, four (later six) office towers, surrounding a 70-story hotel, built on prime downtown riverfront. Ramrodded by Henry Ford II, with the intention of reviving the central city (even though his own company's headquarters had been in Dearborn for years), RenCen enlisted the capital investment of almost every major corporation that did business in, or with, Detroit. Unfortunately, its completion spurred a further emptying of existing downtown office towers, as many prestigious tenants fled their 1920s quarters for the visibility (and higher rents) of RenCen.

The 1980s brought the same office building excesses to Detroit as to other large cities, primarily to the north and northwest. Several new towers went up downtown, fueled by outside investors, but again, they filled their space at the expense of older buildings, which found that their lower square-foot costs could not match the attractiveness of the new buildings.

There didn't seem to be a supply of second-tier tenants to replace the departing occupants.

While there was a wealth of work for architects and builders designing and erecting new office buildings on every major suburban artery, it was a disaster for owners and investors faced with empty buildings and crushing mortgage payments. Some highways seemed lined with "For sale or lease" signs. By the middle of the l990s, the overwhelming vacancies were slowly being worked off, but only at the cost of non-existent new construction. A few large factories in excellent locations were remodeled for new office and apartment use, but most fell to the wrecking ball.

STATE OF THE ART

One of the largest architecture, engineering and construction projects of the mid-1990s is the Chrysler Technology Center and executive offices complex in Auburn Hills, with a consequent wind-down of the company's facilities in Highland Park, some of which dated back to the earliest days of the auto industry.

Two huge state-of-the-art automobile factories went up within the city proper, the first in many years: the General Motors Poletown plant that straddled the Detroit-Hamtramck border, and the Chrysler Jefferson North assembly plant. Both were made possible by governmental assemblage of hundreds of acres of residential, commercial and light industrial land through the city's power of eminent domain, federal grants, and significant tax abatements on the new facilities. Both companies demolished older existing plants, leaving large gaps in several older neighborhoods.

Approaching the second half of the '90s and the end of the century, the construction picture was mixed. "Exurban" subdivisions of expensive homes were being built on former farmland, but little housing could be built for the family of modest means on expensive land, with high cost labor and materials, and increasing interest rates. Office and commercial building had to await the working-off of the glut of the 1980s, and industrial construction was mainly additions to existing plants and a few small new factories. Population of the city was shrinking, and the growth of the state was not keeping up with that of the nation. A $100 million federal grant for a midcity "Empowerment Zone" promised to aid the resurgence of the core, but its full effect would not be felt until the turn of the century.

Fred Pelham (Reginald Larrie collection)

Minorities Make Their Mark

They created a legacy, not just buildings & inventions

by Reginald Larrie

As early engineering graduates from the University of Michigan, and later in their professional prominence, Fred B. Pelham and Cornelius H. Henderson were outstanding role models for contemporary minorities in the engineering, construction and design fields.

Fred B. Pelham was the youngest of seven children born to Robert and Frances Pelham in Detroit soon after they arrived from their home state of Virginia. Once he was old enough, Fred Pelham started working with his father who was recognized throughout Wayne County for his skill as a masonry contractor. Pelham developed a talent for building and, after completing high school, was sent to the University of Michigan to major in engineering. With his high aptitude in mathematics and good study habits, Pelham graduated at the head of his class in 1887.

Greatly impressed with Pelham's knowledge and skill, a Professor Greene recommended him for a job with the Michigan Central Railroad. Immediately, he was hired as an assistant civil engineer. During his years with the "road," Pelham designed and built some twenty bridges in various locations across the state. One bridge (a "skew arch" type) still stands in the small town of Dexter and is the only one of its kind in the country. Pelham held his position with the Michigan Central until he passed away in 1895 after a short illness. He was a member of the Michigan Engineering Society and a teacher in the Bethel A.M.E. Church in Detroit.

Another graduate from the University of Michigan in those early days was Cornelius Henderson, who was born in December 1887 on Detroit's east side. His father was the Rev. James M. Henderson, pastor of Ebenezer African Methodist Episcopal Church at that time. Cornelius, or "Cornie" as he was affectionately called by his many friends and associates, graduated from Michigan in 1911. He was viewed by some instructors as the most outstanding student in their classes. One professor told his senior engineering class that he considered Henderson his "best student."

After graduation Henderson was not as fortunate as Fred and could not find a job for several months because — then as now — most companies wanted someone with experience. However, one day he encountered a friend from U of M and learned about a position at the Canadian Bridge Works in Walkerville, Ontario. Cornelius applied for the position and was hired instantly. Thus began a relationship that lasted some thirty years.

At the Bridge Works, Henderson became involved in the design and building of two great structures — the Ambassador Bridge and the Detroit- Windsor Tunnel, which have operated since 1929. He was a founding member of the National Technical Society.

His great-grandson, Cornelius Henderson IV, graduated from the U-M College of Engineering with a B.S. in mechanical engineering and applied mechanics in 1995.

Continuing a tradition established a century ago by Pelham and Henderson, today another African American graduate from the University of Michigan is making headlines in the engineering field.

HAD A DREAM

His name is Howard F. Sims and from the time he was a young man, he wanted to be an architect. Born in Detroit on July 25, 1933, Sims attended public schools and afterwards began his career in the military with a construction unit of the U.S. Navy. Sims recalled, "The Seabees paid me all of $90 a month but most important was the experience I received. It paid off with two

The Dexter, Mich. bridge designed by Fred Pelham. (Reginald Larrie collection)

years' working credit towards my state architectural license."

With his degree from University of Michigan and a state license, he started his business by opening a design office on Main Street in Ann Arbor in 1963. Today, Howard Sims is president of Sims/Varner and Associates Inc., with offices in the Penobscot Building in downtown Detroit. (Architect Harold R. Varner, one-time director of the Detroit Housing Commission under former Mayor Roman S. Gibbs, joined the firm in 1973 and has been the executive vice-president since 1976.)

Sims and Varner contracts have included low-cost housing, rehabilitation of apartments, expansions of schools (Boynton Middle School) and construction of a moderate-income housing complex, Sheridan Place.

After finishing Sheridan Place, the firm was commissioned to expand Cobo Hall's Conference/Exhibition Center. This renovation was completed in 1989 at a cost of $225 million. The construction nearly doubled the hall's size to over 2 million square feet and gave Detroit one of the largest convention centers in the nation.

To date, Sims has directed the planning and design of over $2 billion worth of construction in Michigan and is still building more. Even with this kind of working record, Sims has not forgotten his community. He has served the Engineering Society of Detroit for a number of years as a fellow and twice as a trustee.

He is also active in many community organizations and is a life member of several civic groups.

With his firm's latest project, the construction of the new Museum of African American History well under way, Howard F. Sims still is a long way from closing down his drawing board.

Willie Hobbs Moore was the first African-American woman to graduate from the U-M College of Engineering (BSE EE '58, MSE EE '61)as well as to earn a doctorate in physics there (1972).

In 1977, Ford Motor Company recruited Moore away from her research/lecturer position in the U-M Department of Physics. As a senior quality associate at Ford's World Headquarters, she helped implement the manufacturing quality control concepts of Professor Genichi Taguchi at Ford.

Moore's early professional experience included positions at Bendix, KMS Industries, Barnes Engineering, Datamax Corporation, and Sensor Dynamics. In 1991, she was listed among the "100 most promising black women in Corporate America" by Ebony magazine.

A native of Atlantic City, N.J., she died March 14, 1994 at the age of 59.

FIRST MINORITY FELLOW

Charles S. Davis, the 55-year-old president and founder of Charles S. Davis and Associates engineering consulting firm when he passed away early in 1995, was not a native Detroiter. Davis came to Detroit when he was hired by Ford Motor Company after earning his doctorate in civil engineering from the University of Missouri-Rolla in 1972.

But Davis was recognized for local distinction when he was chosen as the first minority member of the College of Fellows of the Engineering Society of Detroit in 1980.

This distinguished engineer was a native of Dallas, Texas, and received his bachelor's degree in civil engineering from Prairie View A&M University in Texas. Subsequently he obtained a master's, also in the civil engineering discipline, from the University of Washington in Seattle.

After working as a structural engineer at Ford, Davis formed his own consulting firm in 1978. In addition to ESD, he was a member of the Society of

Dr. Willie Hobbs Moore (U-M College of Engineering)

Engineers and Applied Scientists, the Society of Experimental Stress Analysis, the Consulting Engineers Council and several engineering honorary organizations.

John G. Petty is known as an engineer endowed with the qualities of positive outlook, patience and persistence. He was born Aug. 2, 1938 to John and Frances Petty of Ferndale, Michigan, and later attended Ferndale High School. He continued his education with a degree in mechanical engineering from Lawrence Technological University. Still not content, he later obtained a master's in business management from Indiana Northern University.

In recognition of his technical and management abilities, not to mention many years of struggle and hard work, John G. Petty was promoted to program manager of the "Fox Vehicle Program" for General Dynamics, a major defence contractor in Warren. In this position, he has managed the engineering and manufacturing team developing the "Fox," a multi-faceted armored vehicle for the U.S. Armed Forces. The Fox is a highly mobile, amphibious six-wheel, 19-ton, 600-hp, 500-mile-range unit with crew of three for detection of nuclear and chemical contaminants.

Earlier, Petty was involved in the development and testing of the gas turbine engine which powers the M1 Abrams tank. Petty has not only demonstrated success in industry but has been diligent in the public sector as well. He is a member of the Board of Trustees at Lawrence Tech., active in ESD as vice-president and will soon be the first African American to serve as the organization's president. A true modern leader and role model, Petty has many other community, educational and professional alliances.

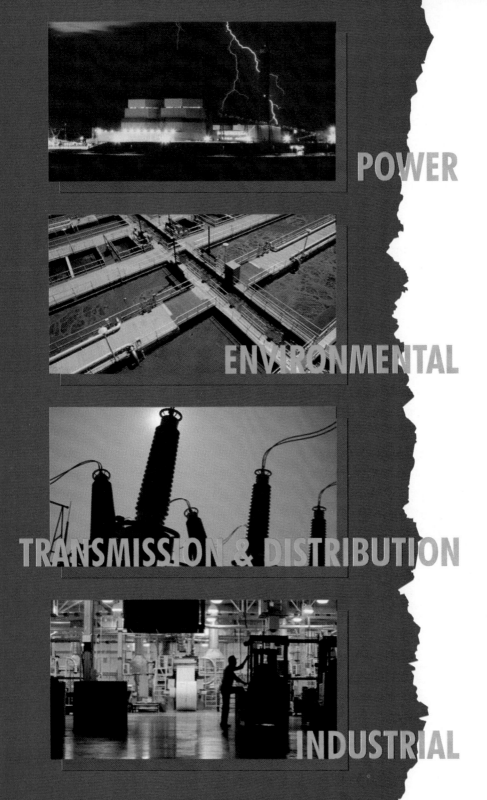

Alias Smith, H&G

*Venerable firm pays tribute to Hinchman & Grylls
as well as to founder — and descendants*

The year was 1853, and a 35-year-old architect was designing a five-story hotel in Sandusky, Ohio. The hotel was the largest building in town so it's reasonable to assume he had been practicing for some time, but no one can find a single document about Sheldon Smith's work before that year. It's almost as though Smith, Hinchman & Grylls' founder had been born a practicing architect!

In 1855, Sheldon Smith moved his practice from Sandusky to Detroit. We don't know why he left Sandusky but Detroit's appeal was plain. The city's explosive growth in manufacturing and commercial enterprises had sparked many new building projects and Smith was eager to design them all.

TOP: *Draftsmen in the Mortimer Smith & Son architectural offices, predecessor of 1995's SH&G, worked by gas light in this photo, ca. 1883. (SH&G)*

SH&G was the architect/engineer for the award-winning Focus:HOPE Center for Advanced Technologies, an adaptive re-use of an abandoned inner-city facility for training residents in a state-of the art manufacturing environment. (SH&G)

His son, Mortimer, born in 1840, virtually grew up amid Smith's architectural practice. At age 21 he joined his father as principal of the firm. Twenty-two years later, Smith's grandson Fred was admitted to partnership. The Smiths' work became so well known that they were selected to design the Michigan Pavilion at the 1893 Chicago World's Fair, only the second worldwide exhibition held in the U.S. There Fred Smith met a young electrical engineer, Theodore Hinchman. Ten years later, in 1903, Fred joined his practice with Ted Hinchman and Ted's partner Henry Field, becoming Field, Hinchman & Smith. The new firm thrived but four years later Field resigned to try his fortunes in the growing automotive industry. He was replaced by H. J. Maxwell Grylls, a British-trained architect with a fine reputation in residential work. The year was 1907. And the company name changed to Smith, Hinchman & Grylls (SH&G).

The firm was blessed with continuity; all three men continued in the same positions until they died more than three decades later. The firm was blessed with opportunity; SH&G was right on the ground floor of the new automobile industry and the support industries that grew up around it. A new factory designed for Henry Ford on Piquette Avenue in Detroit in 1904 led to plants for the "Dodge Boys" as well as residences for the new automobile elite. Castle-like Meadowbrook Hall, now on the Oakland University campus at Rochester, designed for Mathilda Dodge and Alfred Wilson in 1920, represents a high point in Tudor Revival residential design probably never to be seen again.

The Golden '20s heralded an intense building boom in Detroit. And SH&G was hip-deep in it all. The Buhl Building, the Guardian and the Penobscot, all SH&G designs, towered over factories and churches and defined the city's financial district. The firm helped shape Detroit and served companies like Ford, GM, Chrysler, Michigan Bell Telephone, J.L. Hudson, and Hiram Walker. (Its first major job after the depression was a distillery in Peoria.) It also has a long association with institutions of higher learning. The Horace H. Rackham School of Graduate Studies, built in 1938 on the Ann Arbor campus of the University of Michigan, symbolized the university's increasingly important graduate work. For SH&G, it marked the end of an era.

Decades of Design Excellence

Post-war expansion put SH&G into a whole new arena of building opportunities. A one-of-a-kind technical center for General Motors in Warren, Michigan brought together the best minds in design and engineering in the late forties. The National Institutes of Health and a Lunar Receiving Laboratory for NASA took the firm into uncharted technical areas in the '50s and '60s. A petrochemical works for Pasa Petroquimica in Argentina and a medical school and hospital, part of a master plan for Yonsei University in South Korea, allowed SH&G to re-enter a global market.

In the '70s, the firm designed a new corporate headquarters for retail giant S.S. Kresge, now Kmart Corp.; a $100 million-plus complex for Detroit's Harper-Grace Hospitals; a huge new central terminal complex for William B. Hartsfield Atlanta International Airport; a new R&D center for General Electric, and an assembly facility for Deere & Company.

The '80s brought a new series of design and engineering challenges: The Beckman Institute for Advanced Science and Technology at the University of Illinois, named 1990 Lab of the Year by R&D Magazine; a major replacement addition for Methodist Hospital in Indianapolis; new high-security headquarters for the Defense Intelligence Agency and the Central Intelligence Agency; a $100 million assembly plant for Diamond Star Motors in Bloomington, Illinois, and at home, the renovation of the landmark Wayne County

Building, a new engineering school at the University of Michigan and a new R&D center for chemical giant BASF Corporation.

In the '90s, SH&G has aided Detroit's new mayor, Dennis Archer, guiding his 35-member Land Use Task Force to recommend guidelines for citywide land use and development well into the 21st century. The firm upholds its role in making Detroit a better place to live and work, and looks forward to implementing the recommendations outlined in A Framework for Action: A Report for Community Discussion.

Nationally Known

On the building front, new SH&G designs grace the East and the West: a biomedical research center for BASF Bioresearch in Worcester, Mass.; an award-winning health care facility for the historic U.S. Soldiers' and Airmen's Home in Washington, D.C.; Phoenix (Ariz.) Civic Plaza, site of the 1995 NBA All-Star jam session; a family medicine center for Mayo Clinic in Scottsdale; a million-square-foot design-build project for Hughes Missile System Co. in Tucson; a materials science laboratory for Los Alamos National Laboratories in New Mexico and an award-winning recreational complex for Western Michigan University.

Before the decade ends, we will see completion of a $229-million hospital for the Veterans Administration in Detroit, a new corporate headquarters for Chrysler Corporation in Auburn Hills, a $206-million medical center for the U.S. Army at Fort Bragg, North Carolina; new diesel engine test cell facilities for Caterpillar Tractor in Shanghai and Moscow, and a new office complex for the FBI's Criminal Justice Information Services Division in Clarksburg, W. Va.

These kinds of projects make a difference in the lives of architects and engineers. They have allowed SH&G to grow and prosper for over 140 years.

Throughout its history, SH&G reached for many talented professionals to solve diverse design and engineering problems. Architects like Wirt Rowland, Minoru Yamasaki, Sigmund Blum and Ralph Youngren made their mark on the firm and the design profession. Like his predecessors, William Jay Hartman AIA, SH&G's current director of design, continues to add to SH&G's rich design legacy with the nationally acclaimed Focus: HOPE Center for Advanced Technologies in Detroit and the Francois-Xavier Bagnoud Aerospace Engineering Building at the University of Michigan. Hartman is now guiding a group of talented young designers into the 21st century. Their designs will add yet another chapter to SH&G's history — and allow it to thrive for many years to come.

GO AHEAD AND DREAM...

DYLARK® RESINS CAN MAKE IT A REALITY

DYLARK RESINS

Design freedom.
Recyclability.
Proven value.

More than 75% of today's soft instrument panels are made with DYLARK resins. That's proven performance.

DYLARK resins provide an extraordinary blend of heat resistance, dimensional stability and processing qualities that allows design engineers to create cost-effective interiors with excellent energy management characteristics.

As new demands are placed on designers, DYLARK resins are proving themselves again and again for instrument panels, consoles, headliners, etc. They're ideal for lightweighting, reducing heterogeneousness, and DFA/DFM. And, active recycling programs are in place today for DYLARK resins.

Your creative work doesn't have to be a solo effort. We can be with you, from concept to finished component, bringing technical help, analyzing part structure and performance, and finding ways to reduce tooling, processing, and assembly costs through part consolidation.

ARCO Chemical Company
Headquarters: 1-800-345-0252

Our Detroit Automotive
Center: 810-353-6730

ARCO Chemical

Rendering of the Integrated Technology Instructional Center, under construction in 1995 on the University of Michigan's North Campus. (Albert Kahn Associates)

A Kahn-do Kind of Guy

Top projects, honors reinforced architect's role in industrial design

Albert Kahn was born in 1869 near Frankfurt in Rhaunen, Germany, the oldest of six children. His childhood was spent in the Grand Duchy of Luxembourg, where he attended school. His father, Joseph, a rabbi and teacher by profession, believed he might find better opportunities to support his growing family in the New World. The presence of relatives in Detroit prompted him to take his family there in 1880.

Although early in life young Albert had evidenced musical talent, poverty had prevented him from making much headway in this direction. He then dreamed of becoming an artist. In Detroit he found a job in the office of the architect John Scott, but in time he became discouraged at his lack of progress. Hearing of his plight, the sculptor Julius Melchers permitted him to attend his Sunday morning drawing classes free of charge. The discovery that the young pupil was partially colorblind precluded his becoming an artist, so Melchers found a place for him in the architectural office of Mason and Rice.

In 1891, at the age of 22, he won a Traveling Scholarship, an award of $500 from the magazine "American Architect and Building News." The money financed a nine-month study tour through England and the continent. The last three months of his trip were in the company of Henry Bacon, later to become the architect of the Lincoln Memorial in Washington, D.C. Bacon was so helpful that Kahn referred to that period as his "formal education in architecture." After his travels, he returned to Mason and Rice until 1895 when he established his own firm with George W. Nettleton and Alexander B. Trowbridge, two draftsmen from Mason's office.

Within a few months, Trowbridge accepted a professorship at Cornell University; shortly thereafter Nettleton died. Kahn remained with his own office, amidst a financial depression. He endured and gradually his commissions increased, growing steadily in volume and size.

Early in 1903, Kahn was called upon to design his first significant collegiate structure, the West Engineering Building for The University of Michigan. A few months later the pivotal point of Kahn's entire career occurred when Henry B. Joy, director and president of the Packard Motor Car Company, commissioned Kahn to design a huge 40-acre plant on East Grand Boulevard in Detroit.

The elegant Fisher Building (1928) with its "golden tower" on West Grand Boulevard in the New Center area. Albert Kahn's influence dominated this area even more than SH&G's in the downtown area, with the General Motors Building (1922) and the Albert Kahn Building (1931), formerly known as the New Center Building. (AKA)

The Packard factory was the first large auto plant in Detroit. Through it, Kahn established the cardinal principle which was to pervade his entire organization: to satisfy the requirements of the client while creating quality architecture as well. Combining utility and beauty, the new plant was laid out in the shape of a hollow square, with windows on all sides to admit light.

Between 1903 and 1905, Kahn designed nine buildings of conventional wood-mill construction for Packard. While this type of construction was typical of the day, Kahn did not favor it for heavy industry. It produced a working floor obstructed by columns, and the wood floors soaked with oil were a heavy fire risk despite sprinkler systems.

PLANT DESIGN BREAKTHROUGH

In 1905, however, using steel sash, a

Kahn designed most of buildings in Ford Motor Company's Rouge complex, shown here in a ca. 1955 photo. (Mike Davis collection)

novel English-designed product, and concrete reinforced with an innovative metal bar developed by his brother Julius, Kahn designed his then unprecedented and now historically significant tenth building on the Packard site. These innovations provided considerably more natural light and unheard of volumes of space unobstructed by interfering columns while simultaneously reducing the time and money then required for steel construction, the only alternative method known at that time.

Henry B. Joy was not the only man involved in the burgeoning automobile industry in Detroit, however. Henry Ford was looming on the horizon and like Joy, Ford always was looking for ways to improve the efficiency of production. Realizing that his two Detroit plants were inadequate, he purchased a 60-acre tract in suburban Highland Park with the hope of enlarging and consolidating his operations.

He conceived the idea of having an entire plant under one roof with no open courts and no dividing walls. To make his idea a reality, he sought Kahn's assistance, simultaneously establishing a long and prosperous relationship which wed the creative genius of two pragmatic men who changed the course of industrial manufacturing and design. Ultimately, Kahn would design more factories for Ford than for any other manufacturer, and Ford work would be the cornerstone of Kahn's practice in the prosperous '20s.

By 1913, there were more than 860,000 automobiles on the road. The auto industry became an institution which not only revolutionized business but trans-

formed modern life, and Kahn's role in bringing about this revolution and transformation was crucial. His coordination of unprecedented functional efficiency and improved working conditions resulted in an overwhelming stream of commissions from all the auto manufacturers. Over the decades, he did major work for Chrysler, designed more than 150 major buildings for General Motors and over 1,000 important structures for Ford alone.

Other industries watched Kahn's achievements eagerly. Soon commissions followed for food plants, textile mills, cement plants, clothing factories. With the outbreak of World War I, many of Kahn's buildings were turned into plants that supplied the military with vital equipment. Kahn also designed many structures for the government. Examples include Langley Field, Virginia, which introduced the first hangar building ever erected, and many of the Army air fields, naval bases, cantonments and numerous other types of structures. He also was responsible for most of the Ford River Rouge complex as well as many buildings which housed the then growing aircraft industry.

ORGANIZATIONAL MODEL

Albert Kahn's office was the prototype for almost all of today's large- scale architectural organizations. His was among the first to utilize a multi-disciplinary approach that could guarantee and deliver the kind of quality package — by the agreed upon deadline — that was traditionally thought to be possible only from a small office. In 1920, this well-staffed organization

ISN'T IT GOOD TO KNOW THAT SOME THINGS NEVER CHANGE?

Since the turn of the century, Detroit Ball Bearing has seen a lot of change. War, the Great Depression, and man walking on the moon are just a few. Through our challenges and successes, we have grown to be leaders in quality industrial distribution.

Detroit Ball Bearing began nearly 80 years ago in the kitchen of our founder T. B. Moore and his wife, Anna. As Invetech, we have grown to be a leading industrial distributor of bearings and power transmission components. Throughout the years and challenges, we have succeeded because we are guided by innovation, quality, professionalism and service. Simply, we aim to please our customers.

Prepared for our future, Invetech combines the strengths of our resources — people and technology, providing customer satisfaction through inventory technology innovation.

INVETECH COMPANY CORPORATE HEADQUARTERS • 1400 HOWARD STREET • DETROIT, MICHIGAN 48216 • 313-963-6011

Located throughout 19 states with distribution centers in Charlotte, North Carolina; Denver, Colorado; and Detroit, Michigan

numbered 400 and was responsible for more than a million dollars worth of construction work weekly.

While the Kahn office was still growing both in size and complexity, Kahn realized that all too many successful firms built by one individual deteriorate when that individual retires. In 1922, in order to ensure that his organization would endure, Kahn took the key men of the firm — Albert Kahn, Inc. — into the organization as principals. This policy continued, without interruption, until 1940 when it culminated with incorporation of the firm as Albert Kahn Associated Architects and Engineers and the admission of twenty-five key men to ownership in the firm.

Having been responsible for much defense work during World War I, Kahn's office was foremost among those serving American industry when it shifted into high gear in response to the demands of World War II. Notable among buildings of this era were the famed Willow Run Bomber Plant, the Dodge-Chicago Airplane Engine Plant for Chrysler, structures for United Aircraft and its divisions, Wright Aeronautical and Curtiss-Wright, the Glenn L.

Albert Kahn (AKA)

Martin Company, General Motors Corporation, Ford Motor Company, and many others.

In 1938, when his staff numbered nearly 600, Kahn was responsible for 19 percent of all architect-designed industrial buildings in this country and 22 percent of The University of Michigan' s collegiate buildings then in use. That same year, "Architectural Forum" devoted an entire issue to Kahn's efforts, noting that few firms, if any, could open an atlas and spot buildings of their design on all five continents.

Certainly, there were none which could show structures dispersed throughout the entire northern hemisphere as could the firm of Albert Kahn.

Kahn's industrial achievements were as varied as the industries he served. There were silos, steel plants, distilleries, rubber factories, textile mills, filtration plants, smelters, docks, warehouses, creameries, air-

plane plants and foundries. Not all of Kahn's achievements were industrial, however. His versatility encompassed nearly all building categories. Numerous honors were accorded him by fellow architects, competitors, educators and others for his imaginative designs for hospitals, residences, schools, banks, clubs, hotels, theaters, office buildings, newspapers, laboratories and educational facilities.

In addition to countless awards the firm received, Kahn received many distinguished individual honors. In 1933 he was awarded an honorary doctor of laws degree by The University of Michigan; in 1937 he was made a member of the French Legion of Honor; and in 1942, just prior to his death at the age of 72, he received yet another honorary doctorate, this time in Fine Arts, from Syracuse University. Posthumously, he was awarded the coveted Frank P. Brown medal, in 1943, from the Franklin Institute of Pennsylvania "in recognition of his outstanding achievements in the development of industrial architecture."

CONTINUING EXCELLENCE

Albert Kahn Associates, Inc. has grown and prospered throughout the years, and is presently ranked as the 21st largest architectural/engineering firm in the U.S. The firm's 265 employees provide expertise in planning, architecture, and all principal areas of engineering.

The solid foundation of the firm's 100-year history and its innovative contributions to the architectural and engineering design technology have been recognized by a long list of clients who have entrusted the firm with more than 13,000 projects worldwide, adding up to more than $15 billion in unescalated construction costs. Annually, Kahn designs approximately 50 major projects representing $500 million in construction costs.

AKA's projects have brought the firm numerous honors from governmental, civic and professional associations. These include the American Institute of Architects' prestigious "Firm of the Year" award for "consistently producing distinguished architecture through the collaboration of related architectural disciplines."

Recent AKA "greenfield" automotive projects include: Mercedes-Benz's new 1.2-million square-foot Sport Utility Vehicle Production Facility in Tuscaloosa, Ala.; the new 1.3-million square-foot BMW Automotive Assembly Plant in Greenville, S.C.; the 1.5-million square-foot Chrysler Jefferson North Assembly Plant in Detroit; the Nissan Truck Assembly/Stamping facility in Smyrna, Tenn.; and Mack Truck's Assembly facility in Winnsboro, S.C.

The Chrysler Technology Center in Auburn Hills, largest new office and engineering complex built in the industry in nearly 40 years. Construction under the management of Walbridge Aldinger was still under way in 1995, with many other local local design, engineering and construction firms involved in the project. (Walbridge Aldinger/Balthazer Korab)

Builders' 'Guild'

Detroit's boom spins off many construction firms rated as tops

by Martha Hindes

Creating a city's skyline, its neighborhoods, its industrial corridors, took vision for the 20th century pioneers who built Detroit. It was the kind of vision that could turn the pale lines on a blueprint and mounds of brick, cement and steel into palaces of industry or countless comfortable homes.

The men who were drawn to Detroit as it was emerging as the "Motor City" had that kind of vision. Young architects and builders such as George B. Walbridge and Raymond Giffels, Alvin E. Harley and Carl O. Barton had ideas for raising strong structures out of the flat fields that dotted the city's early landscape.

They would make their mark in an industrious race that, in a few years, would add indelible landmarks and a strong, new regime of building designers and engineers marching toward international fame.

In 1924, when Barton, his sister Margaret Rose and an associate, Kins Collins, kicked off the C.O. Barton Company, General Contractors in downtown Detroit, they had an established building trades workforce to draw upon. While operating engineers could command 90 cents an hour, and an electrician got a dollar, common laborers could be hired at half the price, just 50 cents.

Barton, a University of Michigan civil engineering graduate, owned all but two of the initial 1,000 shares

Simplicity is wisdom.

Complexity breeds confusion.

Confusion erodes profitability.

We couldn't make your life any simpler.

The simplicity of one.
One global company can cut through the clutter and give you the systems and components you need.
One global company with two worldwide product groups that can respond to your needs throughout North America, Europe and Asia.
One global company can make your life simpler.
In a manufacturing environment, things can get pretty complicated.
ITT Automotive can give you the simplicity of one.
It's a beautiful concept.

ITT
AUTOMOTIVE
The simplicity of one.

•**Brake and Chassis Systems,** which include vehicle stability, traction control, anti-lock brake and total brake systems; chassis systems; body systems; fluid handling systems; precision die castings; friction material; and Koni shock absorbers.
•**Body and Electrical Systems** which include seat, door, window, wiper and air management systems; motor and actuators; switches and lamps; body hardware; and structural stampings.

of stock in the company begun with $500 in borrowed funds. With the renovation of a Michigan Bell Telephone facility as its first assignment, the fledgling company netted a whopping $1,247 its first year, with more than a half-million dollars of work.

Some eight years earlier, the same kind of independent adventure had been begun by two men: Walbridge, who had come to Detroit from Pennsylvania, and his partner Albert Aldinger, a German-born Canadian. Walbridge's experience as a colonel in the U.S. Army Corps of Engineers would serve him well as the new construction firm of Walbridge-Aldinger first opened its doors.

STRIKING OUT

Those and other far-thinking builders were taking a giant step away from the security of employment with older, established companies to start out on their own. Despite years of despair when the Great Depression ground the industry to a virtual halt following the stock market collapse in 1929, many would survive to contribute some of Detroit's most notable structures and win the city's top praise.

As the area grew from a smattering of modest buildings to its expansive, high-rise present, those homegrown companies were making an impact.

In Walbridge-Aldinger's portfolio from the early 1900s are its work on the Ambassador Bridge, whose graceful span connects Detroit to Windsor; Olympia Stadium; and the Penobscot Building, for years the city's tallest.

Some 225 design and detail assignments by the 60-year-old Giffels Associates firm — known as Giffels and Vallet when it was established in 1925 — helped give rise to the Ford Motor Company's massive Rouge complex in Dearborn. The progressive march of the auto industry would power up that firm and others as the country emerged from the First World War and later the Depression years.

In the 1920s as auto manufacturing was becoming an industrial force, the workers hired to make it run needed housing. John E. Roth, a young engineer, and his brother, William F. Roth, started a construction firm after World War I that ultimately would build 500 houses in Pontiac and Flint for General Motors Corp.'s Modern Housing Corp. division. A 1927 model, the "Winchester," sold for around $7,000. Roth, Inc. and its sister architectural firm, Roth & Associates, now do only industrial and commercial construction, including a specialization in food warehouses.

As the century progressed, so did the contributions of the area's architectural and construction firms that themselves were growing beyond local boundaries.

In the late 1920s, when Carl Barton was deep in debt, Arnold Malow had joined as a partner bringing with him business-saving financial backers. The firm survived to become today's Barton Malow Company. Although its projects have been concentrated in North America, by 1983 it had earned a spot as 25th largest construction firm worldwide.

Its sports facilities such as the Silverdome in Pontiac, Hubert Humphrey Metrodome in Minneapolis and Detroit's Joe Louis Arena have provided homes for American sports teams. Among other Barton Malow projects were Detroit's Children's Hospital, the Detroit Historical Museum and work on the Mayo Clinic in Rochester, Minn.

As with other area firms, auto industry projects (including Chrysler Corporation's three-year-old Jefferson North Assembly Plant) continue to be a mainstay of its business.

MANY HONORS

The builders of Detroit have won wide recognition in the process of raising a city. By 1978, Walbridge-Aldinger found itself recipient of an outstanding achievement award from ESD — The Engineering Society for its work on the IBM Regional Office Building in Southfield. It would become one of many honors the firm has garnered from ESD and other industry organizations. The firm won the bid as construction manager for Chrysler's four-story technical center and, currently, its 15-story administration building.

Two other construction firms have made an impact on the area. Turner Construction, headquartered in New York, established an office in Detroit in 1974 when it won the new Detroit Receiving Hospital contract. It ranked as the area's 11th largest construction firm last year, based on 1994 revenues, and currently is building the city's new African American Museum in the cultural center.

Many area firms have succeeded in outbidding world-class competitors. One of them is the employee-owned BEI Associates, Inc. of Detroit, ranked among the top 10 architectural engineering firms in Michigan by the trade publication Crain's Detroit Business. The company, which started as Hoad Engineers, Inc. in 1953, did all the engineering work on the huge Cobo Conference Center expansion in 1989, then moved some of its efforts to Kuwait for reconstruction following the Gulf War.

Two other Detroit area firms won architectural engineering bids on the huge Chrysler Technical Center complex on 504 acres in Auburn Hills. Those companies — Giffels Associates and Harley Ellington Design, Inc., both of Southfield — have storied histories from Detroit's early days.

After only two years on his own back in 1912, Harley Ellington founder Alvin Harley had wrested a commission away from Albert Kahn, his former employer. The prize, an English Cottage-style mansion, the first of many elegant homes he designed, won him a spot among the city's top architectural firms. Ellington joined him during the low-tide years of the Depression. In 1970, Ralph Pierce and Warren W. Yee also became partners.

Back in the 1930s, after surviving the Depression years, the Giffels architectural/engineering firm had established a specialty in foundry design, then added industrial nuclear facility design after World War II. Its projects include Detroit's original Cobo Hall and the Kennedy Space Center's lunar launch complex. The firm, renamed Giffels Associates, Inc. in 1970, now includes two subsidiaries: Giffels Technologies, Inc., of Southfield, specializing in advanced technology projects; and Giffels Hoyem Basso, Inc., of Troy, architects and engineers.

HIGH-TECH METHODS

The emergence of computer technology has changed the thin blue line on architectural and engineering drawings to the bright gridlines on a video screen. And computer aided design and virtual reality, which allow three-dimensional prototypes to be explored and manipulated from many angles, are fast becoming the state of the art — even old hat — for design firms.

As a latecomer to the local scene just 13 years ago, Dearborn's Ghafari Associates, Inc., under the guidance of founder and owner Yousif Ghafari, has pushed the limits of technology along with acquisitions in its quest to expand worldwide.

The firm, which quickly incorporated computer aided design and engineering techniques, now has representatives in 12 foreign cities. Its international focus has brought it architectural and engineering projects in such places as Brazil, China, Saudi Arabia and Poland. And the purchase of three longstanding rivals (Benjamin, Woodhouse & Guenther; McKenzie, Knuth & Klein; and Eberle M. Smith) during the past six years has expanded the company's base.

Detroit firms, faced with outside competition, have held their own. Southeast Michigan design firms have been well represented in Crain's Detroit Business tally of the top 20 architectural engineering firms doing business statewide based on 1994 revenues.

And builders place highly among the publication's 1994 tally of the top 20 general contractors as well. In that ranking, Barton Malow placed first. Walbridge-Aldinger, which in 1984 had acquired the construction firm of Darin & Armstrong, came in second.

Detroit area builders have a new foundation for their continued success, the adoption of the current "design and build" philosophy of construction that has led to corporate pairings or joint ventures.

That idea of "turnkey" construction — building from idea to completion — spurred Robert DeMattia 18 years ago to found the R.A. DeMattia Co. of Plymouth. DeMattia's foresight put his company about six years ahead of the now widespread trend. The firm, with a specialty in high technology building, sometimes uses joint ventures with outside architectural specialists to meet design requirements beyond the scope of the group's architectural entity, DeMattia & Associates, Inc.

Garages were the first projects of another relative newcomer, the family-owned DiMaria Building Company, Inc., of Novi. Following construction of the precast Federal Aviation Administration (FAA) tower at Detroit Metropolitan Airport in 1991 — at 230 feet, the tallest in the FAA's Great Lakes Region — the firm has expanded to include three divisions covering health care and housing including low-income projects, the industrial/university sector and special projects involving governmental and municipal work. For two years, 1993 and 1994, it has ranked eighth in the Crain's ranking.

While building construction can be a massive enterprise, complete with earth moving blasts and bulldozers, steel haulers and cranes, sometimes the work calls for a delicate touch.

That happened about five years ago for DeMaria, known for its ability to handle projects with special requirements, according to Judy Ferris, client coordinator for the firm.

"We did a renovation of the open heart surgery area of Port Huron Hospital," she recalls. "We put up a partition wall and were doing construction on one side while they were doing open heart surgeries on the other side at the same time."

Electrical Engineering

Detroit Edison lights the way

by Dan Jarvis

TOP: *To meet the public's ever-increasing demand for electricity, new power plants were started by Detroit Edison at Conners Creek on Detroit's east side in 1913 and at Marysville, near Port Huron, in 1921. This photo of the southeast elevation of Conners Creek, showing its landmark "seven sisters" stacks, was taken Sept. 20, 1920. (Detroit Edison)*

The foundation of the world's electric and electronic systems is based on the work of such early pioneers as Volta, Ampere and Ohm, but it was Thomas Alva Edison who made electricity a commercial reality when, on Sept. 4, 1882, he opened the first central generating plant on Pearl Street in New York City. This plant is memorialized at Michigan's Greenfield Village, where its only surviving dynamo is preserved.

The Pearl Street plant housed six direct-current dynamos, each driven at 350 revolutions per minute by a 200-horsepower reciprocating steam engine taking steam from hand-fired, coal burning boilers. Each dynamo had the capacity to power 1,200 incandescent electric lamps. The direct-current, two-wire underground distribution system served customers within a one-quarter mile radius.

The Edison Illuminating Company, a forerunner of The Detroit Edison Company, was founded in 1886, and served downtown Detroit customers within a one-mile radius of its power plant at the intersection of Washington Boulevard and State Street. It was at this plant that Henry Ford in the 1890s served as chief engineer while in his off hours he tinkered with his experimental gas-engined motorcar. Much of this original power plant, known as "Station A," also was preserved by him at Greenfield Village.

Although New York's Pearl Street station operated for eight years with only one three-hour interruption, the success of electricity was by no means a certainty. These early electric plants were small because most of the load was used for lighting and because Thomas Edison's direct-current (DC) distribution system severely limited transmission distances. Soon, however, electric motors were powering elevators and industrial equipment — necessitating the need for increased electric production.

George Westinghouse's successful demonstration of alternating current (AC) with the Niagara Falls-Buffalo transmission line in 1896, and Nikola Tesla's invention of the alternating current poly-phase motor and transformer, laid the foundations for interconnected transmission systems and the industrial electrification of the 20th century.

AC-DC

The first AC plant in Detroit was in service by 1900, and by 1902, AC lines linked some of the downtown power plants. Rotary converters in each plant provided alternate sources of power for the DC distribution system, which improved reliability and reduced the need for duplicate generation.

The Detroit Edison Company was founded in 1903 by the consolidation of several early electric companies in Detroit. The new company responded to the rapidly growing demand for electricity by starting construction of a new plant in February 1903, the first of three built on the Delray site. As originally designed, it housed two of General Electric's new vertical turbine generators rated at 3 megawatts (MW) each and generated power at 4,600 volts.

Fires were lit under the first Delray boiler in August 1904, and additional turbine generators were ordered in 1905 and 1906. Foundations were laid in 1907 for Delray Power House Number 2, for which Detroit Edison ordered 14-MW turbine generators. During the first decade, the Delray plants increased to a rated capacity of 80 MW. To meet the company's ever increasing demand, new plants were started at Conners Creek in 1913, Marysville in 1921, and Trenton Channel in 1924.

The original Trenton Channel Power Plant housed six turbine generators, with a generating capacity of 50 MW each — 17 times the size of the original Delray generators. Two decades after Delray 1 and four decades after Pearl Street, the six units at Trenton Channel produced a total output of 300 MW, a huge power plant at that time.

Trenton Channel, however, was noteworthy for more than its size: rather than being burned on a grate, its coal fuel was pulverized and blown into the boiler furnace much like fuel oil. This made Trenton Channel a true engineering landmark because it was one of only two utility-size installations with pulverized coal firing.

CLEAN AIR APPARENT

Installation of the nation's first utility-size electrostatic precipitators on the Trenton Channel boilers was another engineering landmark. Long before concerns for clean air were voiced, the company installed equipment to remove most of the airborne ash produced by the burning of coal. Detroit Edison has never installed a pulverized coal-fired boiler without the use of electrostatic precipitation technology.

In two decades, the Detroit Edison electrical system had expanded from one which served a number of communities more or less independently to an integrated system about 100 miles long. To tie the power plants and their loads together, the company completed its first 120,000 volt (120-kV) overhead transmission line in 1924, from Marysville through Pontiac and Ann Arbor to the Trenton Channel plant.

The rapidly increasing demand for electricity and resulting demand for new plants stimulated numerous improvements in generating equipment. During the 1920s and 30s, the sizes and ratings of turbine generators increased, as did the pressures and temperatures of steam, and Detroit Edison was in the vanguard of these advances. During the 1930s, the company carried out experimental work on a small turbine generator using steam at 1,000 degrees from an oil-fired super heater. As steam pressures, temperatures and unit sizes continued to increase, so did overall thermal efficiency. Improvements came so quickly that plants became obsolete in little more than a decade.

The increasing demand for electricity — especially reliable electricity — also resulted in many advances in transmission and distribution. Although early substations were operated manually, by 1926 a number of fully automated, unmanned substations of a standard 23-kV design were in service.

DIRECT CURRENT'S DEMISE

Load on the DC distribution system in downtown Detroit, which reached a peak of 32 MW in 1930, was exceeding the system's technological capacity. Conversion to an AC network system began in 1930, and the last of the DC system was finally retired in 1961.

As technology improved, it became possible to serve customers far from urban areas, and electric service was extended to rural customers. By the mid-1930s, more than 60 percent of the farms in the company's service area had electric service. In 1935, Detroit

Service truck of the Edison Illuminating Company prior to the organization of Detroit Edison in 1903. (Detroit Edison)

Edison acquired the Michigan Electric Company, which served the state's eastern Thumb area, expanding the service area to its present 7,600 square miles. To provide additional line capacity for remaining rural customers, the company increased its subtransmission voltages in the Thumb from 24 to 40 kV. Gradually, the entire overhead subtransmission system was converted to 40 kV.

Although both overhead and underground distribution systems were employed, bulk power transmission in urban areas was limited because cables for voltages greater than 24 kV were not available. In 1941 the first 120-kV underground cable was installed from Delray to Warren Station, a distance of seven miles. This cable consisted of three single conductor cables installed in a seven-inch-diameter steel pipe and insulated with nitrogen gas at a pressure of 200 psi.

RISING TO CHALLENGE

World War II and the demands of industrial plants building military equipment severely strained the capacity of the nation's electric utilities. Thanks to conservative design and the courageous decision to rebuild the Conners Creek Power Plant during the Great Depression, Detroit Edison was able to meet the heavy demand of the area's "arsenal of democracy." After the war, economic expansion resulted in almost continuous building of new power plants as the company's load doubled each decade until the 1960s. Plant efficiency improved as steam temperatures of 1,000 degrees became standard and plants changed from banks of boilers and turbines to single boiler/turbine unit systems using superheated steam.

The company installed its first postwar units — also its first 100-MW units — at Trenton Channel and incorporated many technical advances such as hydrogen-cooled generators, a centralized control room and complete AC electrical auxiliary systems, which became feasible because of the strongly interconnected transmission system.

The next landmarks in the steady climb in unit sizes were St. Clair, the largest power plant in the world when Unit 7 was completed, and River Rouge 1 and 2 which, at 260 MW each, were the largest units in the world when they were commissioned. River Rouge 3, at 300 MW, was another world record-setter in its day. However, these soon were overtaken by the Monroe plant with four 750-MW units boasting a total capacity of 3,000 MW. Less than five decades after the company completed its original Trenton Channel plant, Detroit Edison's newest plant was 10 times the size of its predecessor.

With larger unit sizes and a stable economy with modest inflation after World War II, the cost of these plants was maintained in the $150-per-kW to $200-per-kW range, while efficiency and operating conditions continued to be improved. In all, Detroit Edison has commissioned 24 new generating units since the end of World War II and now has a total generating capacity of 10,000 MW. Compare this with the fact that the company's peak load did not exceed 1,000 MW until late in World War II or that the capacity of Fermi 2 alone is 1,140 MW.

By 1939, the scientific advances of the century's first four decades had demonstrated that a controlled nuclear reaction was probably achievable, and development soon was under way in several nations, including Germany, the USSR and Japan.

Nuclear Family

Fortunately the Allied effort succeeded, and the first controlled nuclear chain reaction was achieved by Enrico Fermi, an Italian physicist, at the University of Chicago on Dec. 2, 1942. The atom was first harnessed to generate commercial electricity in the United Kingdom in the early l950s, followed by the first commercial nuclear power plant in the United States at Shippingport, Pa., in 1957. Today, more than 400 nuclear plants operate in 30 nations, supplying about 17 percent of the world's electricity. In the United States, 109 nuclear plants supply about 20 percent of the nation's electricity. In addition, literally hundreds of reactors are in service in four of the world's naval fleets, mostly powering submarines.

Detroit Edison was in the forefront of nuclear power development when it formed a consortium of utilities and other companies to build the Enrico Fermi 1 breeder reactor, which began operation in 1963. The plant life was short because, as an experimental reactor, it could not compete economically in that age of relatively inexpensive fossil fuels.

Low Profile, Low Current

In the electric utility business, developments in transmission and distribution go hand-in-hand with developments in generation. In 1960, the company took a new approach in construction. In substations, Detroit Edison replaced its tall steel lattice girders supporting bus conductors with aluminum stands only 15 feet high. This low-profile design greatly improved the appearance of such installations and made possible the construction of step-down stations in urban areas.

Increases in both load and load density led to the adoption of 13 kV as a distribution voltage in 1960 to augment the 4,800 V systems standard since early in the company's history. In 1968 the same load pressures resulted in the introduction of a new transmission voltage — 345 kV — with completion of the line between the St. Clair Power Plant and the Pontiac Station. Today the 345-kV "outer loop" forms the backbone of the company's transmission system and its interconnections with neighboring utilities.

Voltages are stepped up at power plants before distribution over lines. Since stepping up voltages reduces the amount of current, the practice allows the transfer of large amounts of power over comparable equipment. Voltages are stepped back down at several points along the distribution system to levels that industrial, commercial and residential customers can accept.

To improve the appearance of transmission lines, the company installed steel poles instead of lattice towers to support a 120-kV line between the Warren and Evergreen stations in 1965. At about the same time, line hardware made feasible a more aesthetically pleasing wood pole line for 120-kV transmission. Such lines now are used for subtransmission feeding substations which transform 120 kV to 13 kV.

More recent developments in transmission included installation of the first 345-kV underground cable between Stephens and Caniff stations in 1973 and the installation at several locations of 120-kV switch gear insulated with sulfur hexafluoride gas. Enclosing and insulating this equipment with gas considerably reduces its physical size, making it suitable for installations in limited spaces where appearance is a concern.

Freed to Refocus

Prior to 1950, Detroit Edison engineered and built its own plants, a practice followed by only a few electric utilities. For nearly the next decade, the company performed all engineering until the rapid growth in demand in the late 1960s forced a restructuring whereby many larger projects were contracted to architects and engineers. Thereafter, Detroit Edison's engineering departments concentrated on solving problems, scoping projects, establishing standards and criteria, monitoring design work performed by the architects and engineers, and supporting operating facilities.

This method is very effective because when a problem arises in a plant, Detroit Edison engineers familiar with the plant and its equipment have an advantage over outside contractors. Company engineering specialists know whether similar problems have been experienced in the industry and, more importantly, know whether other Detroit Edison plants have experienced such problems and how those problems were solved. By incorporating the lessons learned over many years of power plant operation into the design and construction of new plants, company experts have significantly improved the reliability and availability of these new plants.

Detroit Edison engineers have always been active on national code committees, and their many contributions can be measured by the widespread acceptance of many company standards including those governing rules for safe boiler operation, power plant control room design and data-base management systems.

In the mid 1960s, Detroit Edison engineers provided expertise during the design and construction of the

Michigan Electric Power Coordination Center in Ann Arbor, which monitors power plant generation, customer demand and the changing economics of buying and selling electricity among utilities. In the mid 1970s, company engineers guided the design and construction of Detroit Edison's System Operations Center (SOC), which monitors, coordinates and directs the day-to day operation of the Detroit Edison electrical system.

During the design review process of the SOC and several other construction projects, company engineers developed unique methods of evaluating systems for reliability, availability and operability. During the design phase of the Belle River Power Plant in the late 1970s and early 1980s, the engineers used new methods and tools — such as operations research, risk analysis, human factors engineering and computer models — to study heat rates, human performance, voltages and unit sizes. As a result, significant improvements have been made to the company's overall power plant system.

REGULATORY RETROFITS

These techniques are not reserved for new plants. Problems in existing plants often pose greater engineering challenges than those encountered in new designs. Detroit Edison has devised solutions to these problems through the blending of experience and the application of new methods and technologies. A good example is the impact on power plants of escalating environmental regulations over the last two decades. Not only have these regulations required the backfitting of large, expensive additional equipment such as dust collectors which had to be fitted within the confines of operating plants, but in many cases these mandated changes resulted in reductions in plant output and efficiency.

Such losses were expected — and experienced — when the company began burning low sulfur coal from Montana in the St. Clair boilers designed for higher-sulfur Eastern coals. Detroit Edison engineers, using the company's own computer models, studied the effects of additional modifications to the boilers. The result was a relatively simple and inexpensive modification which proved the validity of the models and resulted in improved boiler output and availability using low-sulfur coal while reducing the need for supplementary oil. Detroit Edison remains a world leader in the blending of coal to increase efficiency and reduce pollutants.

The company's latest major construction program, which ended when the Fermi 2 nuclear plant reached commercial operation in 1988, put heavy demands on Detroit Edison's engineering and construction departments, its architect-engineers, and its construction contractors. At the peak of the program, an estimated 1,500 engineers and designers and more than 5,000 skilled tradesmen were at work on Belle River, Fermi 2 and various environmental projects. With these projects completed, the company experienced a sharp decline in engineering and construction activity. However, it still oversees work performed by contractors and can perform smaller construction jobs and emergency or special work.

Now one of the largest electric utilities in the nation, Detroit Edison has 8,400 employees serving nearly 2 million customers in its 7,600-square-mile service area, covering 13 counties in southeastern Michigan.

As part of the drive to provide customers with exceptional value, company engineers have solved many complex problems — such as efficient energy usage and conservation of resources — for industrial electrical users. Working closely with companies such as General Motors, Ford, Great Lakes Steel, Scott Paper and Amoco, Detroit Edison engineers have shown ingenuity and commitment in providing superior service and significant savings by going beyond the call of duty to establish positive relationships.

As a result of the company's three-year, $236-million reliability improvement program to strengthen its 27,000-mile transmission and distribution system and customer communications, Detroit Edison now has the second-best service reliability record in the nation. Improvements in overhead-line maintenance and the installation of 100,000 lightning-protection devices will benefit not only larger industrial customers, but residential and commercial customers as well.

As the company's facilities age, continuous improvements will be necessary to improve service reliability and keep existing equipment functioning properly. It is not surprising that Detroit Edison, like many utilities, has no firm plans for building additional generating units. Whatever the future holds, we can be sure that strong, competent engineering organizations will continue to be as important in the future as they have been throughout the company's illustrious past.

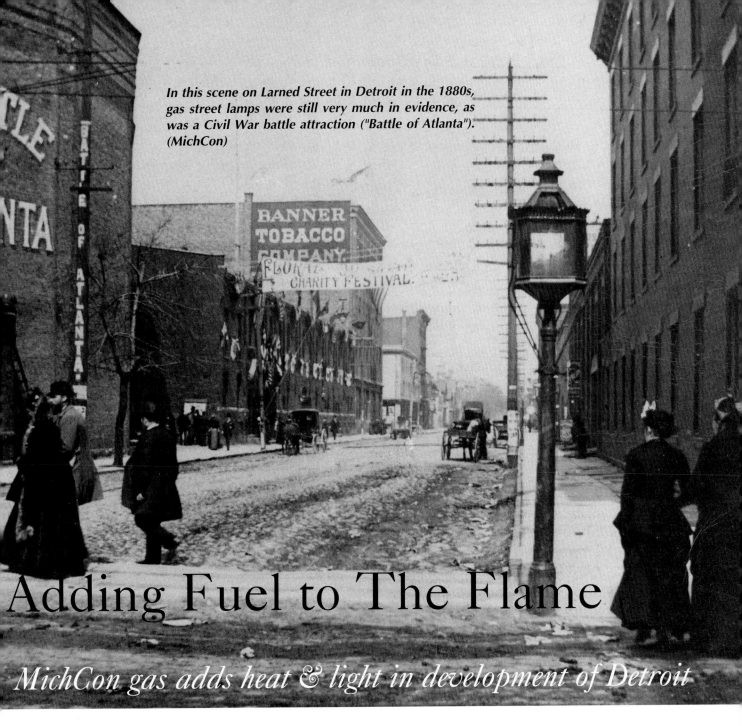

In this scene on Larned Street in Detroit in the 1880s, gas street lamps were still very much in evidence, as was a Civil War battle attraction ("Battle of Atlanta"). (MichCon)

Adding Fuel to The Flame

MichCon gas adds heat & light in development of Detroit

Every fall in Detroit as temperatures drop into the fifties for the first time, there's a click and a whoosh and suddenly the house is warm. What seems so simple to us is actually a process that's been finely honed over nearly 150 years.

Before it was ever used for heating or cooking, gas was used to light our city's streets. The first attempt to bring gas light to Detroit began in September 1848, when 12 of the city's most respected merchants and lawyers organized the City of Detroit Gas Company. More than civic-mindedness, however, was needed to put gas light on the streets. The manufacture and distribution of gas was an expensive and complex enterprise.

Late in the summer of 1849, Lemuel H. Davis, a gas engineer from Philadelphia backed by eastern financiers, came to Detroit to organize and establish the Detroit Gas Light Company. Davis built a gas plant on Woodbridge between Fifth and Sixth Streets. The heart of the plant was a long, low brick building which housed the retorts. Retorts, cast iron containers about eight feet long and two feet wide, were placed in banks of six each over a furnace. The first step in making gas was to start the furnaces and stoke them until the retorts glowed a bright red. Then the workers shoveled soft coal into each retort and banged the lid shut. Inside the airtight retorts the coal, deprived of oxygen, could not burn; in the fearful heat it smoldered away, yielding its light-producing gases. After about six hours, nothing was left in the retorts but coke. The process was known as "destructive distillation" of coal.

The next step was to treat the raw gas removed from the retorts. The hot gas rising from the coal was piped out of the retort house into another building, where a series of condenser pipes cooled it. The cooling process removed water, light oils, tar and some of the ammonia suspended in the gas. "Sulpheretted hydrogen" (hydrogen sulfide) had yet to be removed. This is the gas which emits the characteristic odor of rotten eggs. Containers of "copperas" (iron sulfate) and slaked lime were used to draw out the hydrogen sulfide and most of the other offensive impurities.

The by-products of this process were varied and valuable. If the coal used in the retorts was of a good enough quality, the coke remaining could be sold to iron and steel works or used to heat the furnaces under the retorts. Tar usually sold well, but when its price dropped, it too could be burned in the furnaces. Ammonia salts were later recovered for use in chemical plants or as fertilizer.

What remained were gases that would burn cleanly — carbon monoxide, hydrogen, and methane. This mixture was piped into the "gasometer" (gas holder), which not only stored the gas but also contained a counterweighted iron piston that kept it under proper pressure for distribution.

By June 1851, 53 street lamps had been erected on Woodward, Woodbridge and Jefferson Streets. On Sept. 24, 1851, Detroit was illuminated for the first time by gas manufactured from coal. On that day, a gas light was lit in front of City Hall and the Detroit Gas Light Company was in the business of lighting streets and homes. Later, a Detroit newspaper proclaimed: "GAS! GAS! GAS! The town was illuminated for the first time on Wednesday ... it was a novel and pleasing sight ... the promptitude and energy of the officers of the company is worthy of all praise."

As the desire and need for gas grew and applications expanded beyond lighting to cooking, more gas was needed, as were improvements in manufacturing and distribution. A new plant was built to help accommodate these needs.

The new plant featured the latest improvements in gas production techniques, including fire-brick retorts, which required more time to carbonize the coal than did the old-fashioned variety made of cast iron. But the fire-brick retorts brought the coal up to a higher temperature and thus yielded a better quality of gas.

In 1868, orders went out for one of the new "telescopic" water-seal gas holders, to be built at the company's newly-purchased eastside lot on Chene Street,

between Wright and Franklin. The new holder had a capacity of 520,000 cubic feet, which tripled the company's storage capacity. Thanks to this increase in its storage capacity, the Detroit Gas Light Company was able to increase its total output in 1870 to almost 85 million cubic feet, three times the production of only 10 years earlier. Today, MichCon can store 130 billion cubic feet (Bcf) of gas and its customers annually use in excess of 200 Bcf.

In 1885, natural gas was discovered in western Ohio. The fields were near cities accustomed to paying high rates for manufactured gas, and several companies were formed to utilize the new fuel source. Natural gas pipelines were extended to Toledo, and several businessmen began to consider constructing pipelines all the way to Detroit. In 1888, work began on a 94-mile natural gas pipeline, and on the night of Dec. 31, 1889, employees of the new Michigan Gas Company first lit gas jets erected in Grand Circus and Cass parks. Natural gas lighting had come to Detroit.

While the Ohio supply was close to depletion by 1894, another field was found in Ontario, Canada. So began the second pipeline in Detroit, this one running underneath the Detroit River. A long, sloping trench was dug in the river bank at the foot of Orleans Street, and another across the river on the Canadian side. The wrought iron pipe was assembled on the American side in 200-foot sections. At a signal given by flags, a locomotive stationed on the Canadian side would pull the pipe into the river, stopping just before the pipe sank beneath the surface. Then another section would be added until the pipe was long enough to span the river bottom. Towards the end, the pipe grew so heavy that it took seven locomotives hitched together to produce the power necessary to haul the pipeline up onto the Canadian shore.

Because the supply of natural gas was limited to whatever was found in these Ohio and Ontario fields, manufactured gas supplemented the natural gas. In fact, manufactured gas was one of the most important resources of the Detroit City Gas Company until 1936. That year, natural gas was brought to the city by a pipeline from the gas fields of the Texas Panhandle to gas consumers in Detroit — a distance of 1,200 miles.

The pipeline came above the ground at the property of the Detroit City Gas Company in Melvindale, near River Rouge. A station was built there for the many operations necessary to treat the gas before it could be routed into the city's distributing system. (This facility still exists today.) First, since natural gas is odorless, an odor was added to the gas. The gas also was mixed with a small amount of water and "fogged"

with oil to keep the joints and connections in the mains from drying out. Finally, the pressure under which the gas traveled, about 100 pounds per square inch (psi), had to be reduced to between 10 and 50 pounds, depending on the load conditions and the season of the year. (Today, gas typically travels near 1,000 psi.)

In Detroit, one of the most important industrial uses for gas was to speed the drying of paints and enamels used on automobile parts. In this operation the chassis and body parts were sent on conveyors between banks of gas jets which baked them dry. Other important uses for industrial gas were in forging, japanning, core baking, and for melting some metals. Because of all of the manufacturing, Detroit was using more than its "share" of natural gas. What could be done to ensure the availability of natural gas year-round for manufacturing needs without using up what homeowners needed?

The answer lay underneath the ground. There were some partially depleted gas fields in west central Michigan which might be used as reservoirs to store relatively low-priced gas that could be bought in the summertime for use later on. This is how gas is still stored today, eliminating the need for manufactured gas.

As the industry was developing, so was Michigan Consolidated Gas Company. The Detroit Gas Light Co. (City of Detroit Gas Co.) enjoyed exclusivity until the fall of 1871 when a second company, the Mutual Gas Light Company, began operating in Detroit. A third firm, the Michigan Gas Company, entered the gas business in Detroit in 1889 and four years later bought out the other two firms, creating the Detroit Gas Company. In January of 1898, the company was reorganized and named the Detroit City Gas Company. In the years that followed, many gas and intrastate pipeline operations were acquired by the company, and it grew and prospered as a result. Through the years, the company has operated under several names. Its present name was adopted in 1938; the familiar form, MichCon, was coined in 1982.

Today, MichCon operates a major intrastate pipeline system in Michigan. This system is connected to sources of supply and furnishes gas transportation services to other companies. In addition, MichCon owns and operates storage fields used to supplement its own gas supply during the winter and to store gas for other companies. MichCon also operates a pipeline maintenance and repair service for both emergency and scheduled work situations.

Basically, MichCon is a retailer of natural gas. That means it buys gas from companies that specialize in finding and transporting natural gas, and resells it to our customers. Much of the gas distributed is tapped in the southern part of the U.S. and transported north by pipeline. Primary gas-gathering facilities serving MichCon are in Texas, Oklahoma, Louisiana and nearly 200 miles offshore in the Gulf of Mexico. MichCon also receives gas from suppliers in northern Michigan and Canada.

Although most MichCon customers are concentrated in the metropolitan Detroit area, the company also serves many outstate communities — from Ann Arbor to Grand Rapids, and north from Mt. Pleasant and throughout the Upper Peninsula.

What 12 men started nearly 150 years ago to help make the streets of Detroit safe has become Michigan's largest natural gas distributor, providing safety and comfort to 3.5 million Michigan residents.

Consumers Power Company sign in Royal Oak warning of underground gas line. (Mike Davis)

Power to
The People

*CMS advances reliability of
storage, distribution for rural
energy consumers*

onsumers Power Company, the principal subsidiary of CMS Energy Corporation, originated in 1886 when W. A. Foote formed the Jackson Electric Light Works in Jackson, Mich. The company grew in the late 19th century and early 20th century from separate electric and natural gas utilities, located in several cities in Michigan's Lower Peninsula, acquired by Foote and his successors.

Consumers Power is the fourth largest combination gas-and-electric utility in the United States, serving electricity and natural gas to almost six million people in Michigan. The company was organized in Maine in 1910, and incorporated in Michigan in 1968.

Consumers Power formed NOMECO Oil & Gas Co., a wholly owned oil and gas exploration and production company, in 1967. It formed CMS Cogeneration, now called CMS Generation, in 1986.

A corporate reorganization in 1987 created CMS Energy Corporation, a $3.6 billion (sales) diversified energy company with businesses engaged in electric and natural gas utility operations; independent power production; interstate storage, transmission and marketing of natural gas; oil and gas exploration and production and utility services.

CMS Generation, the independent power unit of CMS Energy, is one of America's top five independent power producers with interests in 4,087 megawatts from 13 operating power plants in the United States, South America and Philippines. CMS Energy was the lead developer of the Midland Cogeneration Venture

(MCV), a 1370-megawatt cogeneration facility in Midland, Michigan, the largest such project in operation in North America.

'LIGHTS IS WORKIN' '

When Consumers Power Co. was in its formative years near the turn of the 20th century, the now-routine process of transmitting electricity from its source to customers was not taken for granted. So in 1899, when the company built a 24-mile transmission link between Kalamazoo and a hydroelectric dam on the Kalamazoo River, Consumers founder Foote did not know what to expect from the new system until a rider on horseback returned from Kalamazoo to the dam yelling, "It's workin'. The lights is workin'."

Now supplying electricity and natural gas to 67 of Michigan's 68 lower peninsula counties, Jackson-based Consumers has always had to be concerned with transmitting power over great distances to reach its far-flung customers. In addition to Michigan's rural regions, the operation serves some of the world's largest industrial giants, including General Motors, Dow Chemical, Upjohn and Kellogg.

Consumers' commitment to serving the farmer was articulated by Wendell Willkie, the 1940 presidential candidate and president of the holding company that owned Consumers Power. He urged the firm to "get out front of the parade" by providing electricity to the farming community. The utility turned on power to its first farm family in 1927, and by October 1949 became the nation's first utility to connect 100,000 farms. That record has never been equaled.

An increased appetite for electricity and appliances led Consumers to develop its Traveling Showroom in 1935. The Showroom transported sample electrical appliances to intrigued rural consumers.

The natural gas end of Consumers' business received a boost in 1941, when the company made its first connection with Panhandle Eastern Pipe Line Company's interstate gas transmission pipeline.

Pushed by Michigan's suburban boom, Consumers' electrical sales surged after World War II to the point where the firm built 19 baseload generating units between 1948 and 1971. The construction boom included the company's first nuclear plant near Charlevoix.

A proposed nuclear plant in Midland, however, almost proved to be the company's financial undoing in the early 1980s. Plagued by massive cost overruns in Midland, Consumers Power Co. halted work on the facility in 1984.

Lines of Communication

The telephone ties together people, enterprises everywhere

Michigan telephone history began just 16 months after Alexander Graham Bell's invention transmitted the first sentence that could be plainly heard and understood.

On July 26, 1877, Bell's "electric speaking telephone" bowed into Michigan history. The scene was Detroit. The Evening News for July 27, 1877, carried this brief account:

"C. C. Reed, telegraph superintendent of the Michigan Central railroad, received yesterday a house

TOP: *Employment as switchboard operators, as shown in this c1950 Michigan Bell photo, was a major source of work for women before the days of equal opportunity and technological advances which obsoleted the equipment. (Ameritech)*

telephone from the inventor A. Graham Bell, of Cambridge, Mass. He proposes to satisfactorily demonstrate its merits by a series of experiments in a few days. It consists of two small boxes in each of which are magnets placed in contact with a diaphragm of sheet iron. There are mouth pieces on the end of each and the sound of the voice, coming in contact with the iron plate and magnet, causes an electric current and carries the sound over the wires. Each box is, of course, placed at the required distance for communication. Besides these boxes there are also wooden attachments, containing only the iron plate or diaphragm, which may be attached at intermediate points on the communication wires. It is claimed that the electricity furnished by the magnets will serve for several of these attachments, and allow other persons to exchange conversation, besides the two persons at the magnet boxes."

The first commercial telephone line in the state was installed in Detroit in Sept. 1877. A single iron wire, strung from the roofs of houses, connected the Frederick Stearns drug store, at Woodward and Jefferson avenues, to the firm's laboratory at Woodbridge and Fifth streets.

A telegraph "throw-switch" and a bell were placed at each end of the line. A battery was installed at one end of the circuit. To make a call, one of the switches was thrown to cut the telephones out of the circuit and bring in the battery and bells. The caller then pushed a button to ring the bells. He then threw the switch back, bringing the telephones into the circuit, and began his conversation.

THROWN FOR A LOOP

Stearns invited the public to use the telephone. A sign in his drug store window read, "Come in and talk over the amazing long distance telephone. Throw your voice almost two miles." (A slight exaggeration in distance.) Accounts of the day report that people flocked in to try the new device, but most were skeptical. They were sure it was a fake and that the voice they heard was someone shouting through a speaking tube from an upper floor.

The second commercial installation in Detroit was for the police department. Here are two of the several newspaper accounts of the event:

"The Bell telephone is now being tried by the police department with a view to its adoption in place of the rickety old telegraph apparatus, which is always out of order. Connections were made with the different police stations, and although the operators were inexperienced the apparatus worked surprisingly well." — Post, Oct. 4, 1877.

"The telephone in the police stations continues to afford visitors and attendants much entertainment." — Free Press, Oct. 6, 1877.

Michigan's first telephone company, the Telephone and Telegraph Construction Company, was incorporated Oct. 31, 1877. In those days, since the switchboard had not yet been invented, communication was possible only between telephones attached to the same line. Telegraph lines often were used to carry telephone messages. In some cases, temporary phone lines were built on top of houses and office buildings. It wasn't long, though, until the telephone company began the first real pole-and-wire telephone lines.

Pioneer linemen drove leisurely in horse-drawn wagons to construction sites, then hoisted the poles and stretched the wire with raw muscle power. The poles were pretty much like the ones used today, except they were longer and not so well protected against decay. Lines, however, were made of soft iron similar to baling wire. For a crude protection against rust, the wire was boiled in oil.

GROUNDED CIRCUITS

A circuit used a single wire. The return connection was made by grounding the line in the earth. Telephone men were pleased when their line paralleled a railroad because they could ground the circuit on a rail and get superior full-metallic transmission.

The telephone was now a permanent part of the Michigan scene. Telephone transmission had proved successful and the first operating Bell telephone company had been formed. But the switchboard had not yet appeared in Michigan. A half-dozen telephone companies in other parts of the country, however, had devised workable switchboards.

On Aug. 5, 1878, Michigan got its first switchboard. It was a crude piece of equipment set up in a small room at the rear of the American District Telegraph Company's office, which occupied the lower floor of a building at the corner of Detroit's Griswold and Congress streets. The 40-story Guardian Building now occupies that site.

The switchboard was manned by several messenger boys of the American District Telegraph Company.

The first directory of telephone customers was published on Sept. 15 of that same year. It had 124 listings. The police department, although listed just once, had ten telephones, bringing the total number of telephones for that time to 133.

Grand Rapids got Michigan's second switchboard on June 1, 1879. Soon, telephone exchanges were established in Bay City and Saginaw. Port Huron citizens, who saw their first telephone in 1878, celebrated the opening of their exchange on Christmas Day, 1879. Other cities with early exchanges were Cadillac, Charlotte, Fenton, Flint, Kalamazoo, Lansing, Manistee, Muskegon, New Haven, Pinckney, Pontiac, Royal Oak, Rockland, St. Clair, Ypsilanti, and Wyandotte.

It was during this period of expanding telephone service that "the voice with a smile" was born. Girls began replacing boys as telephone operators. Early accounts say the boys scurried about in a crazy fashion and that the rowdier ones even swore at subscribers who insisted on quick connections. Grand Rapids is credited with being the first in Michigan to employ women regularly as operators. That city took the step when its boy operators were discovered shooting marbles instead of tending the switchboard. Detroit saw the wisdom of the move and employed its first women operators late in 1879.

The license granted by the national Bell Telephone Company to the Telephone and Telegraph Construction Company of Detroit provided only for the building and operation of exchanges. It did not give the right to construct or operate inter-city or toll lines. An exchange was defined as all territory within a 15-mile radius of the central office.

This upright telephone with a dial for automatically reaching numbers without going through an operator was a major technical advance about 1925. (Ameritech)

Under these terms, the longest telephone lines connected Grosse Pointe and Wyandotte through the Detroit switchboard. And the greatest distance two people could talk was the 24 miles between those two communities. The first international telephone line began service Jan. 20, 1880, between Detroit and Windsor, Ontario.

LONG-DISTANCE LEADERS

Although the western part of the state had started constructing a long distance line from Muskegon via Grand Haven to Grand Rapids, the honor of completing Michigan's first long distance circuit went to the builders of a line from Port Huron to Detroit. This line was looked upon as a marvel of telephone construction. It was then, in 1881, one of the longest toll lines in the world.

As the need arose to link the exchanges in the state, new companies were formed. One was the Michigan Bell Telephone Company, created on Jan. 10, 1881. Within a few years it was combined with the Telephone and Telegraph Construction Company to form the Michigan Telephone Company. From 1882 through 1885, other cities connected to Detroit by telephone included Ypsilanti, Ann Arbor, Toledo, Jackson, Coldwater, Hillsdale, Flint, and Lansing. Saginaw, connected in 1884, boasted the first circuit of copper wire, the conductor that provided the best transmission over long distances.

When Alexander Graham Bell's phone patent expired in January 1894, the door was flung open to unlimited competition in the growing phone industry. It was a period that brought about the adverse conditions that led to the concept of the telephone industry as a natural monopoly.

One of the first non-Bell affiliated companies to spring up in Michigan was an Escanaba firm established in 1894. Another much larger one was Citizen's Telephone Company which opened for business in Grand Rapids in 1896 and operated there and in Kalamazoo, Lansing, Jackson, and other central Michigan communities in 1923.

Duplicate telephone service was common in those early years. Competing companies organized in the same areas meant costly duplicate facilities, and that customers had to subscribe to both companies' services to be able to call all telephones in town. The last example of duplicate telephone service in Michigan ended in 1950, when the Jonesville Cooperative Telephone Company system became a part of the Michigan Bell system.

Though the Bell system ended duplication of telephone facilities, even in 1970 there were more than 70 telephone companies operating in Michigan, most of them small. Several, however, were large companies with thousands of phones and hundreds of employees. The thinking then was that, in place of competition — believed to result in poor service and unnecessary customer expense — just one telephone company would serve a given area, tied in to all other areas where telephone service was available. All phone companies in the state are utilities regulated by the Michigan Public Service Commission. In 1995, the monopoly concept for local telephone service is once again being challenged.

While the 20th century was still in its infancy, the telephone industry grew into a vigorous youngster. By 1906, the Michigan State Telephone Company (later to become Michigan Bell Telephone Company and today's Ameritech) had 200 exchanges and nearly 100,000 subscribers.

That same year, the company's first automobile garage was built in Detroit. Its five types of vehicles were symbolic of the times and place: an "electric cable truck," a "light gasoline wagon" manufactured by the Oldsmobile Company of Lansing, a "heavy gasoline truck" manufactured by the Reliance Motor Car Company of Detroit (predecessor of GMC Truck), a "gasoline delivery wagon" manufactured by Cadillac and a "gasoline runabout," also built by Cadillac.

INSTALLER CIRCUITS

The first motorcycle unit of telephone installers was outfitted in Detroit in 1912. Many early installers made their rounds on bicycles.

Work was begun on the New York to San Francisco telephone line in 1913. It was completed on Jan. 25, 1915. And, within a month the first commercial telephone call from Detroit to San Francisco was made. The three-minute call between two executives of a Detroit automobile firm took 30 minutes to set up and cost $16.70. Today a three-minute call to anywhere in the continental United States can be made for $1 or less — and the call goes through instantly.

Immediately after World War I, telephones grew rapidly in use and popularity. By April of 1919, the new Detroit Headquarters Building of the Michigan State Telephone Company (Michigan Bell) was ready for its first occupants. By August, all administrative offices of the company were in the new seven-story structure at Cass Avenue and State Street. Twelve stories were added in 1928.

When radio broadcasting in Michigan began in 1922, the telephone system was ready to play its part. Telephone wires were used to transmit programs from the studio of WWI (now WWJ) to its transmitters, just as they are today. The first radio station hookup via telephone wires — in 1924 — connected Detroit and 11 other cities for the broadcast of the Republican National Convention from Cleveland. Today the nationwide telephone system provides the coast-to-coast radio and television networks that transmit programs to stations throughout the nation.

Another important milestone in the telephone's development occurred in Detroit in 1923: automatic dial telephone service was introduced by the Michigan State Telephone Company (Michigan Bell). It took until after World War II before dial service became nearly universal.

BELLWETHER EVENT

On Jan. 1, 1924, the Michigan State Telephone Company — formed in 1904 when the old Michigan Telephone Company was reorganized — officially became the Michigan Bell Telephone Company. Renaming of the firm was a symbolic event publicly indicating the company's close association with the other Bell System companies in the nation, a relationship that was to last 60 years.

The first transatlantic call from Michigan to London, England, was made on Feb. 12, 1927.

The first public demonstration of television was April 7, 1927. Bell engineers successfully transmitted the image of Herbert Hoover, then Secretary of Commerce, from Washington, D.C., to New York City. Three months later, the bulky TV apparatus was brought to Detroit for a demonstration. Two-way air-to-ground speech with airplanes, using regular Bell System facilities, began in May, 1929.

Teletypewriter exchange service was born on Nov. 21, 1931.

Development of the coaxial cable system, which has played a major part in long distance telephone development and network television, was announced in the fall of 1936.

When the "fateful 40s" dawned, Michigan Bell's number of telephones had hit the 820,000 mark, 306 central offices were in operation, and customers were making over four million calls each day.

New telephone services and new equipment came onto the scene with astonishing rapidity. Telephone communication between moving vehicles became commonplace. Ship-to-shore service for the Great Lakes ships was inaugurated in 1942. Mobile service, for telephoning from cars and trucks, was introduced in 1946.

In 1948, the nation's telephone system was ready to carry television programs over its coaxial cables and telephone radio-relay networks. These long-distance communications were augmented in the '60s by microwave transmissions and in the '70s by satellites.

The personal communication revolution of the 1990s has been the wireless cellular telephone, shown here being used outside Ford's Wixom Assembly Plant. (Ameritech)

LITTLE DEVICE, BIG NEWS

The big news announcement of 1948, perhaps of the decade, told of the invention of the transistor by a Bell Telephone Laboratories science team. It revolutionized the telephone and electronics industries. Its tiny size, long life, and stingy use of electrical energy opened dozens of new avenues for electronic and communications development. The discoverers of the transistor, physicists William Shockley, John Bardeen, and Walter Brattain, were later awarded the Nobel Prize in Physics.

The 1950s and early '60s brought even more telephone advancements — and at a faster clip. On July 15, 1952, Michigan Bell's two-millionth telephone was installed — just 10 years after the million mark was reached.

A year later, customer dialing of long distance calls — Direct Distance Dialing — was inaugurated. Birmingham customers got the service on Nov. 20, 1953. It was the second DDD installation in the nation. (Englewood, New Jersey, was the first.) DDD gradually spread to other state communities, including Detroit in 1960. Service refinements and new instrumentalities emerged at a rate previously unheard of. Michigan Bell installed its three-millionth telephone in 1961.

During the late '50s and early '60s, the Princess phone, the Bell Chime ringer, home interphone systems, and volume control phones were introduced to Michigan Bell customers and were accepted with enthusiasm. Touch-Tone calling, push-button "dialing," made its debut in 1964.

Electronic Switching Systems (ESS) was introduced in 1966, and handled telephone calls a thousand times faster than conventional equipment. This system, which applied computer programming techniques and electronic technology to the switching process, was hailed by engineers as the greatest advance in switching since dial calling was introduced more than 40 years earlier.

Also in the '60s, special school-to-home telephone installations became available for home-bound students. The artificial larynx, a speaking device for persons left voiceless through accident or disease, was developed by the Bell Telephone Laboratories and offered to the public at cost through the Bell operating companies.

Businessmen were able to step up efficiency and production with the introduction of DataPhone data communications service, speakerphones, the Bellboy

Number, Please ...

*Back when phone numbers were still recited alpha-numerically,
certain exchanges had an A-1 kind of ring to them*

Before the advent of rotary dial telephones — which preceded TouchTone phones and today's dawning voice-actuated devices — it was necessary to make telephone connections by actually talking to a real, live operator every time.

You picked up the receiver — originally the microphone and the earphone were separate — and heard a voice, almost always female, ask "Number please?" In the beginning, the number was short, one, two or three digits.

As the number of telephone lines grew, exchange names were added. These exchange names took on a magic all their own. Frequently they referred to neighborhoods. For instance, downtown Detroit was — and still is — the Woodward exchange; "96" is the same as "WO." Telephone exchange names also carried social panache: Birmingham and Bloomfield Hills were the Midwest exchange and it stood for something.

Glenn Miller, fabled '40s bandleader, made a New York City exchange and number forever famous in the great number "Pennsylvania Six Five Thousand," with its lyrics meaningless to anyone who never had to give a number to an operator.

When the telephone company replaced exchange names with straight numbers in the '60s, again because of huge growth in lines and the need to have additional numbers which didn't depend on words, geographic identity of exchange areas became blurred, then lost.

The disappearance of exchange names coincided with direct-dial of long-distance calls, in which distant exchange names lost their meaning. Area codes, along with the new postal zip codes of the '60s, added another dimension to international geography.

pocket signaling set, and a myriad of other services to satisfy just about any communications need. In 1967, toll-free 800 lines were introduced, which over the years became an incredible boon to marketing and servicing thousands of products and services for millions of customers.

One pair of telephones in 1877 grew to 4.4 million Michigan Bell units by 1969. From those phones — then virtually all owned by the company and merely leased to customers — Michigan callers could reach some 118 million phones throughout North America and 238 million all over the world.

Editor's Note: The preceding "History of the Telephone in Michigan" was adapted from a brochure issued by Michigan Bell about 1969, the most recent company history present-day Ameritech was able to provide. Since that time, the telephone industry both locally and internationally has been revolutionized — organizationally and technically. Following are a few notes on those changes:

- *In 1970 facsimile transmission ("fax") was still in its infancy, used by companies sparingly over expensive, crude machines. Today, every home and office computer has the capacity with inexpensive add-ons to send and receive such communications, and stand-alone fax* machines are even being used in moving vehicles via cellular telephone hookups.

- Nationally, intercity picturephone service began in 1970 and came to Michigan a year later. Used mostly for conferences, it never seems to have reached the promise initially envisioned.

- 1970 also saw introduction by Michigan Bell of many popular custom calling services, such as call waiting, speed calling and three-way calling.

- In 1972, computerized operator technology began to replace plug-in cord switchboards.

- During the '70s, high-speed data links connecting business customers' computer centers were introduced, and all major Michigan cities became equipped with electronic switching systems replacing electromechanical systems.

- In 1983, digital switching equipment began replacing electronic and electromechanical switching centers, which enabled many new services.

- In 1982, the Supreme Court of the United States ordered the breakup of the Bell Telephone monopoly, to take effect Jan. 1, 1984. As a consequence, American Telephone & Telegraph Company (AT&T, the familiar "Ma Bell") was separated from both its manufacturing and local service affiliates. Michigan Bell became part of Ameritech, a regional telephone service company covering Michigan, Ohio, Indiana, Illinois and Wisconsin. AT&T retained its long-distance business, but now had to compete with other carriers such as MCI and Sprint.

- The historic breakup in telephone companies was paralleled by a revolution in telephone equipment available at neighborhood retail and discount stores. First came portable phones, then memory phones and "designer" phones of every description. The telephone company no longer was the sole supplier of telephone equipment or systems.

- Fiber optic transmission began in 1984 with a Detroit-to-Pontiac cable, in which 216 pairs of hair-thin ultra-pure glass strands could carry 16,000 individual telephone conversations. Michigan now has some 400,000 miles of fiber optic cabling.

- The demand for ever more telephone lines was accelerated by pagers, personal data links (including fax/modem) and ultimately cellular phones — initially awkward units to plug into vehicle cigarette lighters and quickly to become lightweight battery-powered units carried in purse or pocket. Cellular phone service in Detroit began in 1986.

- Other new customer calling services introduced in the '80s included answering service provided by the local telephone company, call forwarding, automatic callback, caller identification and emergency 911 systems.

- By 1995, Ameritech Michigan could no longer measure its size by number of telephones, as it had a quarter century earlier. For one thing, it no longer owned the phones. More important, the focus had shifted to telephone lines for the many types of communication besides telephones which had come onstream. At the time Michigan Bell/Ameritech separated from AT&T in 1984, it had about 3.75 million number lines; by the end of 1984, that had grown 25 percent to 4.7 million, not counting some 325,000 additional cellular numbers.

- Ameritech has an investment of $7.5 billion in Michigan today, serving 76 counties with more than 16,000 employes. The company is proposing an expansion into interactive cable television on the one hand, and defending itself against incursions for competitive local service on the other.

Does advertising cost or pay?

In the publishing business we are constantly embroiled in an effort to convince potential advertisers that marketing their services to the exclusive membership of ESD—The Engineering Society is a benefit in the long-term for their business.

This is an extremely competitive industry in which costs per thousand, circulation audits, readership surveys and other factors affect the decisions to advertise. Truly, in the scope of the publishing world, ESD's magazines and publications are dwarfed by their multimillion-dollar competitors.

It is for all of these reasons that we are proud to boast this prestigious list of supporting advertisers who have made this book possible. They have recognized ESD's value to the engineering community and to industry. These advertisers have chosen to show their support as an effort to give something back to the organization which now celebrates 100 years of giving to the engineering community.

And it is because other supporters like these have sponsored ESD publications in the past that we have been able to provide our membership with the best in editorial on research and development and applications-based information for new products. Perhaps that information has helped some of these companies to compete or gain an edge in the industry. So, perhaps too, when we consider what this advertising accomplishes, it pays in many ways far and above mere response-driven marketing.

Advertisers in THE TECHNOLOGY CENTURY

Display Advertisers

3M Automotive
American Iron & Steel Institute
Arco Chemical
ATEQ Corporation
BASF Corporation
Bayer USA, Inc.
Black & Veatch
Bridgestone/Firestone, Inc.
The Budd Company
Chrysler Corporation
Construction Association of Michigan
DANA Corporation
DCT
Delphi Interior & Lighting Systems
Detroit Edison
Dixie Cut Stone
Dupont Company
Ferndale Electric
Franco Public Relations Group
Ford Automotive Components Division
Ford Motor Company
French & Rogers Inc.
Gallagher-Kaiser Corporation
GE Automotive
General Motors
Gibbs Die Casting Corporation
Guardian Industries Corporation
Haden, inc.
Harley Ellington Design
Invetech
ITT Automotive
Jervis B. Webb Company
John E. Green Company
Kelsey-Hayes Group of Companies
Kolene Corporation
Lear Seating Corporation
Limbach
Magna International

Masco Corporation
MBM Fabricators
Michigan Benefit Plans
PPG Industries, Inc.
Progressive Tool & Industries Company
Rockwell International
Seabury & Smith
Textron Automotive Company
Turner Automotive Group
United Technologies Automotive
University of Michigan
Walbridge Aldinger
Wayne State University
Lawrence Technological University
Mechanical Heat & Cold
Michigan Automotive Research Corp.
Motor City Electric
NTH Consultants, Limited
Roofing Industry Promotion Fund
Testing Engineers & Consultants, Inc.
j.a. Versical & Associates

Sponsor Advertisers (pages 247-262)

AP Technoglass
American Society for Quality Control
ACT Laboratories, Inc.
Albert Kahn Associates, Inc.
Amcast Industrial Corporation
American Quality Systems, Inc.
Amerisure Companies
AMP Incorporated
ANR Pipeline Company
Arthur Andersen LLP
ASC Incorporated
ATEQ Corp.
B.D.O. Seidman
Badger/Manley Division of Dynagear Inc.
Balogh T.A.G.

Barton Malow Company
BEI Associates, Inc.
Cambridge Industries
City Management Corporation
Crown Plaza Pontchartrain Hotel
CyberMetrics Corporation
Dale Prentice Company
Dana Corp.
DataMyte Division of Allen-Bradley Co., Inc.
Davis Industries, Inc.
Del Technologies, Inc.
DeMaria Building Company, Inc.
DCT
Donnelly Corporation
Dynagear Inc.
DynaLogic Engineering, Inc.
Earth Tech
Echlin Inc.
Ellis/Naeyaert/Genheimer Associates, Inc.
ENVIRO MATRIX, INC.
EQ -The Environmental Quality Company
Federal-Mogul Corporation
Fendt Builders Supply, Inc.
Findley Industries, Inc.
Follmer,Rudzewicz & Co., P.C.
French & Rogers
Fueudenberg-NOK
H.B. Fuller Automotive Technology Center
Handy & Harmon Automotive Group, Inc.
Hoechst Celanese Corporation
Hubbell, Roth & Clark, Inc.
Hull & Asscoiates, Inc.
INA Bearing Company, Inc.
INTEL
Jarvis B. Webb Company
John E. Green Company
Kantus Corporation
LTV Steel Company

Lacks Enterprises, Inc.
Leckie & Associates, Inc.
Lectron Products Inc.
Libby-Owens-Ford
Link Engineering Company
Lucas Assembly & Test Systems
Lucas Industries Inc.
Markel Corporation
Max Brook Inc.
McDowell & Associates
Means Industries, Inc.
Michelin Automotive Industry Division
Modern Engineering, Inc.
NBD Bank
NAMCO Controls Corp.
Navistar International Transportation Corp.
NSK Corporation
PSG Corrosion Engineering, Inc.
Palmer Equipment Co.
Perceptron, Inc.
Perrin, Fordree & Company, P.C.
Philips Technologies
R.H. Jackson Company
Raybestos Products Company
RMT, Inc.
Schroth Enterprises, Inc.
Siemens Automotive
Smith Instrument, Inc.
Stahls USA, Perathane Coatings
Sullivan Ward Bone Tyler and Asher, P.C.
Tenneco Automotive
The Boomer Company
The Woodbridge Group
Tiodize Co. Inc.
Trico Products Corp.
Wayne Foundry & Stamping Company
ZF Industries, Inc.
Myron Zucker, Inc.

The Flow and The Glow

*Flooding the city in light
and usable water, utilities leave
Detroit flush with success*

by Natalynne Stringer-Williams

TOP: *Construction under way in 1954 at Detroit Water and Sewerage Department's Northeast water pumping station near the Detroit River. (DWSD)*

From Detroit's earliest days, when it was little more than a frontier settlement, Detroiters have had to deal with what later in the 19th century became known as the city's public utilities. Supplying fresh water, disposing of wastewater, and lighting the city's streets have been necessities for Detroit, if not since its founding as a fur trading outpost by Antoine de la Mothe Cadillac nearly 300 years ago, at least from the start of its growth after the War of 1812.

The first of these public utilities, the Water and Sewerage Department, had its beginnings in 1824 when, by Legislative Act, Peter Berthelet erected a wharf on the Detroit River to pump water to the village. Detroit's first water-main system was made of hollowed logs.

Since the city's founding in 1701, the Detroit River had provided an unlimited supply of pure water — indeed, early Detroiters simply dipped their buckets into the river for their daily supply of water. But as the population increased, a more efficient method of distribution was sought. The Berthelet pump proved to be a short-term solution, however. In 1827, the Common Council passed an ordinance granting the exclusive rights to provide water to Detroit's 1,500 citizens. The Wells' Hydraulic Company was formed and charged families $10 per year for water. But the demands of Detroit's ever-increasing population soon proved too great for the Wells' company and, in 1836, the City of

Detroit was forced to purchase it to provide adequate management of water needs.

Detroit's sewerage system began that same year. Vast amounts of untreated sewage (through rainwater runoff from streets, where people commonly emptied their chamber pots) discharged into the Detroit River had become a problem. The first effort, known as "the Grand Sewer" and made of stone and brick, enclosed Savoyard Creek that flowed from Cadillac Square to its outlet near the foot of Third Street (a portion of which remains in service today).

FIRST HAS LASTED

In 1853, the City Charter was amended to include a five-member Board of Water Commissioners that was given special powers and authority to expand operations for the city's increasing demands. The following year, the first city-owned reservoir was excavated, to begin operation three years later. New intake, pumping facilities and reservoirs were added as the city (and, eventually, suburbs) grew. Water meters were first used shortly after water fountains were introduced in 1862. The site of the city's first water plant, Water Works Park on East Jefferson near Belle Isle, was purchased in 1873. It is still in operation today.

Gas lighting (distilled from coal) first appeared in Detroit in 1851 and, because of its cost and unpleasant odor, was mostly used as street lighting until the advent of the Welsbach mantle in 1885, which made gas lights more acceptable in homes and offices.

The city's first electric light was exhibited by Charles J. VanDepoele in 1870. VanDepoele, a designer and carver of wooden altars and church woodwork, held the demonstration in his woodcarving shop. The demonstration preceded the debut of the electric light on Detroit's streets by 13 years. In 1879, there were

Detroit's Water Works Park on East Jefferson in 1879. At left is the original pumping station. Next to it is a surge tower which housed a 124-foot standpipe used until 1894 to equalize pressure in mains leading from the pumping plant. The tower was long used as an east side observation point entered by means of a spiral staircase. By the 1940s, the tower was unsafe and believed unrepairable, but was not demolished until 1962. (DWSD and Burton Historical Collection, Detroit Public Library)

1,760 open gas-jet street lamps in Detroit, which then covered less than 16 square miles. The illumination offered by the gas-jet lamps "served only to make darkness visible," as a local newspaper later stated — and the gas lights operated only on nights when the moon was not shining. This practice was known as "moonlight schedule" lighting.

When Hazen S. Pingree took office as Detroit's mayor in 1890, street lighting was one of his highest priorities. Pingree was instrumental in an amendment to the 1893 Public Lighting Act authorizing Detroit to own and operate its own lighting plant. A Public Lighting Commission (PLC) was established shortly thereafter. The city's first power house was built the following year. The plant had seven hand-fired coal boilers delivering steam to 18 Western Electric Company 50-kilowatt arc dynamos for DC series lighting and three 55-kilowatt Westinghouse "alternators" for AC building lighting. Nearly 212,000 duct-feet of conduit served 2,500 incandescent building lamps within a half-mile circle from City Hall on Campus Martius.

So by 1895, Detroit had in place municipal sewer and water distribution systems, and the city's first municipal lighting plant began operation. On April 1, 1895, at 6:35 p.m., switches were thrown in the city's Atwater Plant (where the Renaissance Center currently stands) lighting 189 carbon-arc lamps. By Oct. 1 of that year, 1,493 arc lamps had been taken over from private companies. The savings to city government amounted to $38.80 per lamp per year (in 1895 dollars). Lighting at the time consisted primarily of 724 crane- and 500 tower-mounted lights ranging from 100 to 165 feet in height and carrying 3 to 6 lamps each. By 1900, Detroit's PLC had 2,067 arc lamps and 81 employees. The first power substation was established in the basement of Western High School in 1901.

TYPHOID SCARE

The early 1900s brought a population explosion as Detroit became the automobile center of the world. In 1910, a treaty was signed creating an International Joint Commission to eliminate pollution of waters between the United States and Canada. By that time, some 439 miles of the lateral sewer system and 194 miles of trunk sewer were completed. However, raw sewage was still being discharged into the river at various points and the typhoid epidemic of 1912 proved that the water was unsafe for consumption. In 1913, the Fairview Sewer was constructed from the city limits to the Fairview pumping station to divert sewage flow. Disinfection by adding chlorine began in 1916.

By 1920, people seeking employment in Detroit's automobile industry increased the population of the city to nearly one million. The demand for public service utilities increased accordingly and between 1925 and 1928, the portion of the Detroit River Interceptor from the Fairview station to the Central City was built and construction was begun on the Northwest Interceptor in the Rouge River Valley District. Also in the late 1920s, the expanding Water Department moved into the 23-story skyscraper in downtown Detroit where it is housed currently.

In April 1920, the people of Detroit voted for a municipally-owned street railway system. The additional power needs required that a new power station be built. The Morrell Street Plant went on line in 1927. During the 1920s, 14,732 new street lights were added making a total of 27,171 in service by the end of the decade.

On Oct. 7, 1930, Mayor Frank Murphy formally dedicated the Morrell Street Station, re-naming it in honor of the recently deceased General Superintendent Frank Mistersky. The Mistersky Power Plant changed the focus of public lighting to substation improvements and expanded power for the street railway system. However, the Depression stymied other expansions to the street railway system until a 1937 Public Works Administration grant provided funding and renewed vigor for the project.

The Depression also postponed or curtailed the activities of the city's other utilities. Although parts of the East and West Jefferson Avenue Relief Sewers were constructed in 1930, further improvements were delayed until 1935. Despite the Board of Water Commissioners' authorization for construction of the Springwells Water Plant in 1924, it did not go into full operation until 1935.

The Detroit River Interceptor, Oakwood-Northwest Interceptor, and the Wastewater Treatment Plant were completed in 1940. Detroit and its suburban communities were served by the Wastewater Treatment Plant, which provided primary sewage treatment — removal and incineration of settleable solids and disinfection by chlorination of plant effluent. Designed to serve more than two million, the plant, with additions, would ultimately serve four million.

On July 1, 1940, the Public Lighting Commission assumed responsibility for the operation and maintenance of Detroit's traffic signal system. Expansion and distribution of public lighting slowed, then stopped, following the outbreak of World War II and Detroit's role as the "Arsenal of Democracy."

TURBO-LENT TIMES

The 1950s opened with Detroit's rapid expansion and the development of suburban communities. The United States enjoyed an era of unparalleled prosperity and Detroit's industrial community benefited enormously. The city's lighting system increased its capacity to meet Detroit's needs. In 1950, and again in 1958, the Mistersky Power Plant added a turbo generator (both have since been converted from coal-fired to fuel oil and are still in service).

Although two sedimentation tanks were added to Detroit's sewage treatment system in 1954, it soon became evident that additional treatment was necessary. In 1957, a $33 million program was launched to provide improved treatment methods and expand the sewage treatment service area.

Newer, smaller communities discovered it was more economical to tie into Detroit's system rather than build their own water treatment facilities. With the completion of its Northeast Plant on East Eight Mile Road in 1956, Detroit's water treatment system expanded to serve more than two million customers in 45 communities. Three years later, in 1959, the addition of the Springwells Plant in Dearborn brought a host of technological improvements that, again, increased the system's capacity to meet the needs of more than three million citizens.

The Police and Fire Departments' communications were placed under the Public Lighting Commission's supervision in 1957 and, by 1959, the Commission built four new substations and major modifications were made to seven existing substations to maintain 58,830 street and alley lights.

Trolley-bus service and the Department of Street Railways were terminated in the 1960s. The decade also brought the last major technological change in the city's water system when the coal-powered, steam-fired, high-lift pumps at Water Works Park were electrified. As the John C. Lodge and Walter P. Chrysler Freeways neared completion, the Public Lighting Commission also entered the freeway-lighting business. The additional freeway load and major relighting programs caused the Commission to switch from incandescent to mercury and sodium-vapor lights in 1964 — improving Detroit's street lighting and reducing costs. That year the Detroit Department of Water Supply took over operation of the Southwest Water Treatment Station from the Wayne County Road Commission, increasing Detroit's water system capacity to more than 1,000 million gallons per day (mgd).

As a preventive measure, in 1966, the Board of Water Commissioners sought an alternate water source remote from the Detroit River intake's location. Lake Huron became that source and the Lake Huron Water Plant began pumping water in 1974, bringing the system's capacity to 1,500 mgd. The Lake Huron Water Treatment Station water is pumped nearly five miles from the lake to the plant in Fort Gratiot near Port Huron.

Improvements were also made in the 1970s to the city's sewage treatment facilities to increase solids removal, reduce phosphorus discharge, and eliminate practically all bacterial pollution, improving water quality levels in the Detroit River and Lake Erie. The Detroit Water and Sewer Department (DWSD) continues work in 1995 on a mammoth construction program of advanced treatment facilities at the treatment plant, including final clarifiers, sludge-handling facilities, additional primary tanks, site improvements, extensive electrical renovations as well as general plant improvements.

In 1974, the Public Lighting Commission became the Public Lighting Department (PLD) under the city's new City Charter. Plans were begun on Unit #7, designed to increase the department's capacity by 60,000 kilowatts. In response to the oil crisis of the mid-1970s when energy costs doubled, a more efficient gas turbine generator (GT #1) was put on line. Originally intended for peak load demands, the GT soon approached full-time operation. Freeway lights in the PLD system were reduced in 1983 when the MDOT assumed the cost of maintenance of Detroit's freeways.

As of 1995, Public Lighting, one of the nation's oldest municipally owned electric utilities, operates 86,750 street and alley lights; 1276 traffic signals, 30 substations, more than 5,751 miles of cable, and services more than 100 customers at 1,400 locations.

DWSD's 2,800 employees now serve close to four million water customers in 120 communities in eight counties of Southeast Michigan with a range of 630 to 700 million gallons of water per day. Sewerage treatment services are provided to 75 communities constituting 35 percent of the state's population.

Construction in 1974 of "malfunction junction," (as it was known by mobile CB radio operators at the time), the intersection of M-39/Southfield Freeway and I-96/Jeffries Freeway. (The Detroit News) OPPOSITE: Woodward Avenue in 1909, looking north between Six and Seven Mile, showing what roads were like before paving. Interurban tracks between Detroit, Royal Oak and points north ran along the west side. The nation's first mile of concrete highway was completed here soon after this photo was taken. (The Detroit News)

All Roads Lead to ...

Prosperity, if paved with more than just good intentions

by Louis Mleczko

E arle and Rogers, two names that sound more like a comedy team than pioneers in highway construction, blended their political and engineering talents 100 years ago to help move Michigan transportation into the 20th century.

Horatio S. "Good Roads" Earle and Frank F. "Lift Michigan Out of the Mud" Rogers, along with other road building pioneers such as Edward N. Hines, Roy D. Chapin, Murray D. Van Wagoner and John C. Mackie, were instrumental in transforming Michigan roads from the dirt and wood plank horse trails of 1895 to the modern interstate freeways and surface highways we take for granted in 1995.

In 1895, horse-pulled wagons and carriages were the only "vehicles" on Michigan's estimated 60,000 miles of public roads. Of that total, only 245 miles were improved with either gravel or macadam — a mixture of crushed stone and tar compacted over sand. The rest were merely dirt trails or pathways that often were impassable because of flooding and mud.

The half-dozen wood-plank, privately-owned toll roads weren't much better, and the public was clamoring for an end to paying tolls.

In 1895, the federal government recorded only four automobile registrations nationwide as Washington created an Office of Road Inquiry. And it was bicycles not autos that prompted the development of a modern road system.

Earle was a bicycle enthusiast who, like most bicyclists, discovered that most public roads were unfit to ride on at any speed. In 1899, Earle was elected chief consul for the Michigan chapter of the League of American Wheelmen, a grass roots organization representing bicycle owners.

'GOOD ROAD' GUYS

In 1900, the Wheelmen held the first International Road Congress in Port Huron, and it was there that other "good road" advocates like Hines of Wayne County and Rogers of St. Clair County discussed various road improvement strategies.

The Wheelmen decided political changes were necessary, and later that year, Earle, running as a Republican from Detroit, was elected to the state Senate as a good roads advocate.

By 1903, Earle and his backers were successful in getting the Legislature to create a state highway department and commissioner. Farmers, who dominated state politics, were against the plans, calling the owners of autos those "buzz wagon idiots" that terrorized horses and livestock with their noisy, smelly machines.

The good roads alliance prevailed, and in 1905, Gov. Fred Warner appointed Earle as Michigan's first highway commissioner. Earle then picked Rogers — an 1883 civil engineering graduate from Michigan State College — as the department's first chief engineer.

Earle and Rogers, armed with a first-year budget of $20,000, embarked on plans to build a statewide system of roads, and they advocated using trained civil engineers to develop minimum specifications for roads. That same year, 2,958 autos were registered in Michigan. The license fee was 50 cents.

The fledgling highway department's first road project was a one-mile improved gravel road near Cass City. Two years later, Earle and Rogers proposed tunneling under the Straits of Mackinac, and when told of poor soil conditions there, they said a straits bridge was feasible and needed to connect Michigan's two peninsulas.

AUTO IMPETUS

Meanwhile, the development of the auto

industry in Michigan provided still more incentives for better roads.

In 1901, Roy D. Chapin, a 21-year-old University of Michigan dropout, was hired to drive a one-cylinder Oldsmobile from Detroit to New York so that Ransom Olds could show off his vehicle at a Manhattan auto show. No one had ever driven a car between the two cities purposefully before.

It took Chapin more than seven days to make the 800-mile trip through southern Ontario, upstate New York and the Hudson River valley. He averaged only 14 miles per hour on mostly dirt horse trails. From Buffalo to Albany, Chapin used the Erie Canal towpaths since there were no usable roads along that route.

"The only roads I found were a few miles of macadam near Rhinecliff on the Hudson," Chapin said upon arriving in New York City. "The rest was everything — sand, dust, gravel, ruts."

Chapin, who later was longtime president of the Hudson Motor Car Co., became an influential good roads advocate. Other motorists formed groups, such as the 1902 founding of the Automobile Club of Detroit, to lobby for better highways. Auto manufacturers joined with this auto club to form the Michigan State Good Roads Association in 1906.

In 1909, Earle and Rogers teamed up with Edward Hines, who was picked as Wayne County's first road commissioner, to achieve an engineering milestone in highway construction: the nation's first mile of concrete highway.

MODERN MARVEL

With the help of a $1,000 state grant, Hines direct-

a "superhighway network" for the tri-county area, featuring 21 major roads with 204-foot right-of-ways. Among the roads eventually built from that plan were: Eight Mile, Southfield, Fort, Michigan, Grand River, Northwestern, Schoolcraft, Woodward, Mound, Gratiot and Telegraph.

Michigan added a 2-cent per gallon fuel tax in 1925 to help pay for road improvements, and by 1930, the state's 1.4 million registered vehicles could travel on 2,000 miles of paved roads, including the first paved highway between Detroit and Chicago.

The Great Depression during the 1930s saw Michigan's road programs dwindle in size and scope. In 1933, Murray D. Van Wagoner, a young civil engineer from Oakland County, was elected highway commissioner and began studies for a statewide system of highways and Detroit's future freeway system. Van Wagoner made a detailed study for a bridge at the Straits of Mackinac. The estimated cost was $30 million, and a 4,000 foot-causeway was built from St. Ignace as a prelude to building a bridge.

In 1940, Van Wagoner became the only highway commissioner to be elected governor of Michigan, and with the outbreak of World War II, he launched plans for the nation's first depressed urban freeway, the Davison in Highland Park. The first rural freeway that linked Detroit's western suburbs with the Ford bomber plant at Willow Run was also built.

The Davison freeway replaced a surface road that ranged in width from 60 to 120 feet and was jammed with 24,000 vehicles daily — many destined for Ford's huge Highland Park plant. Started in July 1941, the six-lane, 1.3-mile freeway was built for just $3.6 million and opened to traffic in November 1942. The 10-inch concrete pavement was built without reinforcing steel because of war shortages, but because of superior materials and construction techniques, the original Davison pavement remains in use 53 years later. Today, state highway officials plan to rebuild the outdated Davison: replacement cost, $70 million.

The Willow Run freeway was started in July 1942 with a $5 million federal grant and orders to complete it as quickly as possible so that 42,000 workers could commute more easily to the Ford plant that was churning out B-24 bombers.

"'Now' was the password from Washington," recalled Keith Bagley, a state project engineer. "That

ed Wayne County to build the revolutionary concrete road on Woodward between McNichols and Seven Mile in Detroit. It featured a concrete slab five inches thick and 18 feet wide that rested on a gravel-and-clay base. Road builders from as far away as France, South Africa and New Zealand visited Detroit to marvel at the new roadway, which cost Wayne County $13,354.

From there, road improvements and highway technology grew quickly to keep up with the explosive growth of the auto industry. In 1912, the University of Michigan created the nation's first highway testing lab and hired its first instructor in highway engineering.

In 1913, the Michigan Legislature created the state trunkline road system with Rogers becoming highway commissioner. The new law called for the construction of 3,000 miles of state highways with the state reimbursing local communities for road costs. Rogers lobbied for a gas and weight tax on all vehicles as well as authorizing the highway department to plant trees along state roads.

In 1911, the Wayne County Road Commission invented the painted center line to separate traffic. Detroit erected the first stop sign, and the green-yellow-red traffic light sequence was conceived by Detroit police officer William Potts.

Congress in 1916 passed the first federal road aid legislation, and by 1919, good roads advocates convinced Michigan voters to approve a $50 million bond-financing plan for state roads. Rogers embarked on building 3,600 miles of state trunklines and in 1923 launched state ferry service across the Straits of Mackinac.

GRAND PLAN

In 1925, Wayne County's Hines unveiled plans for

meant pouring concrete in freezing weather without reinforcing steel."

FIRST FREEWAY

On Jan. 31, 1943, the first 11.5 miles of the world's first limited-access freeway opened to traffic, and by March 1945, the four-lane highway was extended to Wyoming road in Detroit. It is now part of I-94.

After World War II, plans for building Detroit's freeway network were revived with Washington pledging $56 million to begin construction of the Edsel Ford and John C. Lodge freeways. Thirteen years and $184 million later, the last 4.9 miles of the Ford was opened to traffic from Conner to Vernier on the city's east side.

The 1950s saw efforts to build new toll roads in Michigan, but the 1956 passage of the Interstate Highway Act saw Michigan take the national lead in freeway construction. The lobby for toll roads vanished.

In 1957, newly-elected Highway Commissioner John C. Mackie, a civil engineer from Flint, launched $700 million worth of bond-financed freeway work. That year also saw completion of the Mackinac Bridge — a $100 million, four-year project that fulfilled the dream first envisioned by Earle and Rogers and later by Van Wagoner. At the time, the 3,700-foot suspension portion of the bridge, designed by David B. Steinman, was the longest suspension span, anchorage to anchorage, in the world.

Between 1960 and 1970, nearly 1,000 miles of interstate freeway were built in Michigan — an average of a mile every few days. Michigan eventually built 1,241 miles of freeway — 90 percent financed by federal funds from gas taxes, at a total cost of $4 billion.

The last section of interstate, a four-mile stretch of I-69 near Lansing, opened in October 1992.

The 1990s saw Michigan's trunkline system at 9,550 miles while all roads in the state totaled 118,000 miles. Some 7.5 million vehicles now travel an estimated 81.5 billion miles annually. State transportation taxes now generate $1.3 billion each year for all highway programs — a far cry from Horatio Earle's first budget of $20,000.

As bicycle enthusiasts, Earle, Rogers and Hines would be pleased to know that there is now also a paved, 26-mile bike-path paralleling I-275 in western Wayne County.

Steam ruled all transportation in 1895. Long distance land travel for passengers and cargo was dominated by railroads like this 1890s passenger train. (CN North America/Grand Trunk) OPPOSITE: This photo dramatically illustrates changes to rail transport by 1995. CN diesel locomotives pull high-cube cargo cars out of the recently enlarged Port Huron-Sarnia international tunnel. (CN North America/Grand Trunk)

Making Tracks

Railroads, though they've declined, were essential lifelines to industry's development

by Bob Cosgrove

To look at where Detroit's railroad's stood in 1895, let's first go back to their beginnings. The first two Detroit railroads to lay track did so in 1837, the same year Michigan was admitted to the union. This was just seven years after the Baltimore & Ohio operated the first American train.

Of the two Detroit roads, the privately funded Detroit & Pontiac, completed in 1843, was significant only locally, although it became the basis for the 20th century commuter line.

More important was the state-owned Central of Michigan, originally projected to run from Detroit west to St. Joseph on Lake Michigan. It resulted from one of the first acts of the new Legislature, which authorized an ambitious $5 million "internal improvements" public works plan calling for three railroads and two canals. Only two of the railroads and none of the canals would be completed.

Known as the Michigan Central Railroad after 1846, when it was sold by the state to eastern financial interests, by 1852 the MCRR line through Ann Arbor, Jackson, Battle Creek, Kalamazoo and Niles linked Detroit with Lake Michigan and Chicago. Detroit lacked a rail connection to the east until 1854 when the broad-gauge Great Western Railway of Canada arrived in Windsor. In 1876 Commodore Cornelius Vanderbilt acquired the Michigan Central for his New York Central empire along with the Canada Southern

Railway crossing Ontario between Buffalo and Windsor.

A major impediment to Detroit's development of east-west rail traffic was the long-unbridgeable Detroit River. The crossing required railroad car ferries and transit became difficult-to-impossible during the ice-bound winter months. And the most direct route east to Buffalo ran through Canada with consequent border-crossing problems. For this reason Detroit became more an important terminal than a major rail hub like Chicago, St. Louis or Toledo.

With Due Dispatch

In 1895, when the Engineering Society of Detroit was founded, trains were dispatched by written train orders communicated to station agents by telegraph. Orders were handed to train crews or passed up "on the fly." Telephone dispatching wouldn't become common until the 1950s. Within 25 years the manual semaphore block signals of 1895 would be replaced with the automatic searchlight and color light signals operated by relays using electric track-detection circuits. In the 1960s the use of radio communication with train crews became prevalent. While not replacing automatic block signals, telephonic dispatching, or even train orders, radio has greatly enhanced operation.

Gone by 1895 was the scene typical of 1895 suburban Detroit depots, when the station agent doubled as a Western Union telegraph operator, ticket seller and handler of light perishable freight such as eggs, milk cans and fresh produce bound for the city. Heavier freight was handled at freight stations, either part of the passenger station, or a separate building with its own freight agent.

Also obvious were the steam locomotive changes after 1895. Over the years until their demise around 1960, these living machines became greater in size, number of wheels, weight and tractive effort. Hidden was their greater efficiency and performance achieved through higher boiler pressures, superheated steam, automatic stokers, feed-water pumps replacing injectors, poppet valve throttles, nickel-steel boilers, mechanical lubricators, and air-operated reversing gear. One hundred years ago not all railroads even used locomotives. The 1872 Detroit Transit railway was horse powered into the 1890s.

Losing Steam

The most obvious change on American railroads was the complete conversion from steam to diesel power in a short 20-year period between 1940 and 1960. General Motors led this revolution.

There had been modest diesel locomotive development of light switching units in the 1920s by American Locomotive in consortium with General Electric and Ingersoll-Rand. GM got into railroad motive power in 1930 by purchasing the Electro-Motive Corporation and its motive power supplier, the Winton Engine Company, both of Cleveland. EMC was a producer of gasoline and diesel self-propelled rail cars and Winton was a continuation of auto pioneer Alexander Winton's business. It was the Winton 201A diesel engine, further modified for heavy rail use by GM's Charles W. Kettering, which enabled GM to open the diesel road locomotive market in 1939 which it captured within 10 years.

The styling of General Motors Electro-Motive Division diesel locomotives in the 1930s can be credited to GM Styling Vice President Harley Earl's automobile design studio in Detroit's New Center, although EMD locomotives were engineered and erected in LaGrange, Illinois and London, Ontario.

Besides this major motive power development, during the first decades of the 20th century the new automobile industry brought the need for more track, marshalling yards and better operating efficiency. In its 1912 annual report, General Motors complained about having to store a million dollars worth of sold cars on city streets during the winter of 1911-1912 due to the inability of the railroads to ship them.

To alleviate such chaos, a number of belt lines circling the different parts of the city were built or expanded to feed the main trunk railroads. These included lines jointly owned by major railroads such the Detroit Terminal of 1905. New team track yards servicing less-than-carload shipments were spotted throughout the city. Large classification yards for sorting and routing cars were built or rebuilt. Rail routes throughout the city were equipped with new signals and interlocking plants to rapidly transfer trains from one route to another. The larger automobile manufacturers located their plants on existing rail lines.

This general upgrading of the physical plant included many highway grade separations (bridges) completed between 1905 and 1914 from Fort Street near Delray to Milwaukee Junction just beyond the present day I-75 and I-94 freeway interchange a mile north of downtown.

BRANCHING OUT

In 1920 Detroit's bottleneck of railroad yards and few east-west lines led Henry Ford to become a railroad owner. The Rouge Short Cut Canal was being dredged to let large Great Lakes ships into Ford's River Rouge complex. This waterway required new rail and highway bridges.

The financially strapped Detroit, Toledo & Ironton Railroad could ill afford its bridge and asked Ford to back its bridge bonds. Instead, Ford bought the forlorn railroad outright for $5 million. The line, known informally as the "Milkweed Central," meandered its way from Ironton on the Ohio River to Detroit. Immediately he applied the lessons learned in his automobile plants. The equivalent of the $5 day was instituted, work hours were reduced and all employees were expected to be clean shaven and sober on and off the job. Locomotives were rebuilt and improved upon at the Rouge, new rolling stock ordered, line improvements made and stations along the route were fixed up, painted and beautifully landscaped.

Ford used the rail traffic of his automobile plants to transform the DT&I's traffic pattern from northbound mine products to south-bound manufactured goods. In Ohio the DT&I connected with major east-west main lines lacking a Detroit access, including the Baltimore & Ohio, Nickel Plate Road, and the Erie.

In 1925, in an almost "Tom Swift" type adventure, Ford developed what was then the world's largest electric locomotive. He intended to electrify the DT&I from his Rouge plant all the way to Ironton, some 300 miles. To this day you can see Ford's enormous concrete arches from I-94 at Oakwood Boulevard and from Allen Road. Weighing 30 tons each, the arches were made at the Rouge concrete plant for the wires of the overhead catenary.

The state-of-the-art engineering for this locomotive was laid out in general concept by Ford engineers and detailed by the Westinghouse company, which supplied the electrical components. The engine was assembled in Ford's Highland Park power house before being tested at the Rouge.

The new Rouge Plant power house supplied the DT&I with single phase, 22,000-volt alternating current (AC) at 25 cycles. This current is easier to transmit over long distances than direct current (DC), which requires substations about every 10 miles. In effect, Ford put the substations on his locomotive.

The locomotive was comprised of two 58-foot units which could be operated separately. Each unit had a 2,000-KVA step-down transformer connected to the 1,675-KW motor-generator set to produce DC for its eight axles each powered by a 670-volt, 225-hp traction motor. It would be 30 years before such an AC line-transmission/DC traction-motor concept was employed by other U.S. railroads.

WORKHORSE

Performance was astonishing. Designed to operate up to 35 mph, the maximum tractive effort (TE) was 250,000 lbs. The largest Mallet compound steam locomotives of that day produced only 150,000 lbs TE. With only one locomotive and the DT&I's relatively low traffic density, the electric operation cost almost twice as much as steam. It ended in 1930 when the twin-unit locomotive was damaged in a derailment.

Ford gave up railroading before electrification was completed beyond Flat Rock. In 1929 he sold the DT&I to Pennsylvania Railroad interests. Ford was used to operating his automobile business without government interference and became disgusted when the Interstate Commerce Commission fined him for not charging demurrage (holding charges) on his own freight cars.

Detroit's railroad engineering landmark of the first half of the 20th century is the Michigan Central's Detroit River Tunnel connecting with Windsor. Begun in 1906, it officially opened Oct. 16, 1910. (This was not the first sub-aqueous rail tunnel in the world, an engineering feat that belongs to the Grand Trunk's 1891 tunnel between Port Huron, Michigan, and Sarnia, Ontario, replaced in 1994 by a new tube.)

Concrete sections of the MCRR's twin-tube tunnel were built in St. Clair, Mich. shipyards, floated to Detroit, and precisely sunk in a trench dredged across

the river bottom. The sections were pumped out and track laid. The tunnel used electric locomotives supplied with commercially purchased power by third rail until 1956, when they were replaced by diesels.

The Detroit rail tunnel is still in daily operation. In April 1994 the present owners, Canadian National and CP Rail, completed enlarging the tunnel to permit passage of tall trains of stacked containers. This was done by lowering the floor of the north tube. The improvement terminated the Detroit River railroad car-ferry service, which had begun in 1867.

A later phase of the original Detroit River Tunnel project was the Michigan Central Depot. Sitting 2,600 feet from the tunnel's west portal, its 15-story office tower remains a landmark on Detroit's industrial west side. Begun in 1912, it opened in 1913. The depot was in Amtrak use until 1987. In forlornly vandalized condition today, hoping for some adaptive use, the large Beaux Arts style station was a major project in a decade which saw many huge buildings go up around Detroit. The architects were a consortium of the firms of Reed & Stem of St. Paul, Minn. and Warren & Wetmore of New York City. These firms also designed New York City's Grand Central Terminal, which also opened in 1913. The MC Depot handled more than 100 passenger trains daily in its halcyon days up to and during World War II.

From 1895 until 1971, when Amtrak was formed, Detroit had three major passenger stations. Besides Michigan Central Depot, there was the 1866 Brush Street Station on the site of today's Renaissance Center and the Fort Street Union Depot on Fort at 2nd, now the site of Wayne County Community College. After using temporary buildings near the MC Depot for seven years, in 1994 Amtrak built a new station on Baltimore at Woodward Avenue just southeast of the General Motors Building.

ROAD-BASED RIVAL

Beginning in the 1970s many of the rail improvements resulting from the early days of the automobile industry have been downscaled or even abandoned. The primary cause has been the flexibility of motor trucks, which better than rail can serve present day "just in time" inventory control methods. The federal government's program of interstate highways begun in the late '50s was another significant factor. U.S. railroads by the 1970s enjoyed only 35 percent of inter-city freight traffic, which had been at 50 percent in the 1950s and almost 100 percent in 1895.

But that downward trend is changing. The 1980 Staggers Act deregulated railroads and truck lines. The railroads have made a significant comeback with intermodal transportation, which combines the advantages of trucks and rail. Intermodal includes TOFC (trailer on flat car) or "piggyback" and COFC (truck-trailer-sized containers shipped on special flat cars, usually stacked two high). The containers are loaded and unloaded from rail cars by crane onto flat-bed trailers or trailer-truck wheels. Intermodal is a concept which goes back to the 1920s and earlier, but has only blossomed in the last 25 years.

Detroit is the northern terminus of a unique Norfolk Southern train called the "Road Railer." Rather than railroad cars, this train consists of specially engineered NS "Triple Crown" standard highway trailers, which include a set of railroad wheels that lower when the trailer rides the rails.

In 1995 CSX Transportation plans to begin using what they call "The Iron Highway" between Chicago and Detroit. This is a train of low platform flat cars, upon which highway trailers will be backed. What is unique is that the flat cars will be permanently coupled and self-propelled. Each end of the train in its final form will have a control car rather than a locomotive. With cabs on either end, the train is bi-directional and need not be turned.

In the last 100 years only one thing about Detroit's railroads has remained unchanged. The gauge between the rails is still 4 feet-8 1/2 inches.

Common Carriage

Commuter transport parallels Detroit's demographics

by Bob Cosgrove

D etroit's commuter transportation system began in 1863 with small enclosed streetcars drawn by one or two horses at a time when most streets were unpaved. Electric-powered streetcars began replacing the horse cars in 1886 and by 1895 the last of the equine four-wheel vehicles was retired.

By 1892 almost all of the city's many streetcar lines were under control of the privately owned Detroit Citizens Street Railway. In 1895 the city under Mayor Hazen Pingree built some street rail lines to be operated by the privately owned Detroit Electric Railway on the condition 3-cent fares be offered in competition with the Citizens company's 6-cent fares.

Citizens acquired the Detroit Electric Railway in 1900, the combination being renamed Detroit United Railways (DUR), and until 1921 Detroit's street railway operations were entirely privately owned. That year, culminating a 32-year campaign for public ownership and backed by a referendum vote, Mayor James Couzens — despite his Ford Motor Company background — began to build new streetcar lines for the city to operate.

Detroit became the first large U.S. city to own its transportation system. The DUR turned its streetcars and trackage over to the city's Department of Street Railways for nearly $20 million. The DSR ended up with 1,457 cars, 363 track miles, 4,000 employees, 10 car houses and two shops capable of complete rebuilding and even new car construction.

In 1925 the DSR inaugurated its first permanent motor bus line. By 1928 the DSR had 38 bus lines. Buses proved a faster and less costly way of extending service than streetcars, although with perhaps higher operating costs. In 1974 the DSR became the Detroit Department of Transportation known as "D-DOT."

Inter-city light electric railways had their Michigan beginnings in 1894. Better known as "interurbans," inter-city electric railroads were an extension of the overhead-wire electric streetcar traction concept.

Electric railways had spectacular growth throughout the midwest in the decade 1895-1910. Using heavier equipment than streetcars, but smaller and lighter than those of steam railroads, the interurbans were a hybrid between streetcars and railroad coaches of the day. In the short six-year period between 1894 and

OPPOSITE: *As this map from a 1916 timetable demonstrates, Southeast Michigan enjoyed an extensive network of electric interurban lines from 1900 to the 1930s. (Jack Schramm collection)*

1900 all of Detroit's 19 interurban lines had been organized.

By 1907 the DUR controlled Detroit and Windsor street railways and all of southeastern Michigan's interurban lines. Detroiters could travel on the DUR to Toledo where, through Ohio electric roads, connections were made to Cleveland, Columbus, Cincinnati and beyond as well as to major cities in Indiana.

'ACCEPTABLE COMFORT'

In Michigan the DUR's lines connected Detroit with other interurban lines to Ann Arbor, Jackson, Lansing, Battle Creek, Kalamazoo, Grand Rapids, Pontiac, Flint, Bay City, Saginaw, Mount Clemens, Port Huron and the many small towns in between. Fares were considerably less expensive than steam railroads. The usually one- and sometimes two-car trains were cinderless and provided acceptable comfort levels at speeds up to 60 mph on frequent schedules. In addition to passengers they carried light freight such as milk cans and packages.

By 1903 Detroit had become the largest interurban railway center in the world, an honor the emerging Motor City relinquished to Indianapolis in 1910. However, Detroit remained an important interurban hub well into the 1920s. For example, in 1919 over 360 cars per day routinely used the DUR's downtown interurban station on Jefferson at Bates, now the site of the City-County Building.

Throughout 1920s interurban railways around the country began to fail in the face of automobile, bus and truck competition. The DUR was forced into bankruptcy in 1925 by its bus supplier, the Yellow Coach company, now General Motors Truck and Coach. It emerged from bankruptcy in 1928 as the Eastern Michigan Railways. By then most of the DUR's original interurban lines had been abandoned or converted to bus and truck service.

The DUR's demise resulted from more than just automotive competition and the cost of maintaining its own right of way. Within Detroit it suffered from lower-cost competitive streetcar fares from its own former system, now municipally operated, and expiring franchises under which trackage reverted to city ownership. Attacks on private ownership, including controlling fares, had gone on continuously since of the Pingree days of the 1890s. With the constant threat of being taken over, the DUR postponed investing in track maintenance and upgrading as well as new equipment and rebuilding.

The long-term harassing of rail transit companies across the U.S. was short-sighted. It denied a reasonable return on investment, prevented capital improvements, caused system abandonments and ultimately forced governments into the public transportation business. The question was whether transit company managements could serve the public interest well at the same time they were beholden to their stock- and bond-holders to maximize investment return. While other utilities survived to continue under private ownership, most public transportation companies, be they bus or rail, did not.

DRIVEN OUT OF BUSINESS

However, the interurbans' short 40-year lifetime in Detroit, as elsewhere, was most adversely affected by the popularity, convenience and flexibility of the automobile. In the two decades 1905-1925 the growth in cars, trucks and buses led to the demand for better roads.

Most interurban rail lines had been laid before the automobile age, down city streets and alongside the crude unpaved highways of that day. Interurban companies spent minimal funds on basic railroad civil engineering. Unlike steam railroads, most interurbans lacked stone-ballasted roadbed and followed the undulating lay of the land to keep costs low.

Few Detroit reminders of the interurban era survive. You can still seen the bridge abutments of the DUR's Detroit, Monroe & Toledo Shore Line Railway over the Huron River at Flat Rock from southbound Interstate 75. South of Flat Rock, I-75 was built in the 1960s over the old interurban right of way. Other interurban rights of way have been used locally for high-voltage transmission routes such as Electric Avenue in Ecorse and River Rouge. A onetime DUR powerplant stands on Grand River at Orchard Lake Road in Farmington.

Not that the interurban died without a fight. For instance, in a 1927 attempt to join the cutting edge of transportation technology, the Eastern Michigan System ran advertisements showing Ford tri-motor airplanes to promote their interurban and bus service which coordinated with airline schedules for persons traveling from Ann Arbor, Port Huron and Detroit to the Ford Airport in Dearborn.

Beginning in the 1920s DUR interurban lines were replaced by motor buses. Some of these bus lines later became Great Lakes Greyhound, Lake Shore Coach, Dearborn Coach, etc. By the early 1970s these companies or their successors had been merged into the Southeastern Michigan Transportation Authority (SEMTA). Formed in 1967, SEMTA was intended to provide public transportation in Livingston, Macomb,

The flexibility and low operating costs of buses quickly doomed electric streetcars and interurbans in the 1920s. Shown here is a 1925 GMC Truck & Coach Model Y "highway coach." (General Motors)

Monroe, Oakland, St. Clair, Washtenaw and Wayne counties. In 1989 SEMTA became the Suburban Mobility Authority for Regional Transportation (SMART). While talks continue of merging SMART and Detroit's D-DOT, these counties still lack a central mass transportation plan and the mechanism to properly fund it.

The development of the motor bus — led by General Motors in the period after World War I — had significant impact on city and suburban electric railways, especially GM's efficient diesel-powered buses marketed from the late 1930s.

Although DSR's total trackage had increased to 533 miles by 1930, it was down to 426 at the beginning of World War II. Detroit streetcars had their greatest ridership ever during the gas-rationing war days of 1942-1945. By 1952, the DSR's trackage was reduced to 65 miles with just four lines: Woodward, Gratiot, East Jefferson and Michigan. Service ended April 8, 1956, and

183 of the 185 streamlined PCC (Railway Presidents' Conference Committee) design streetcars were refurbished by the DSR shops and sold that year to Mexico City.

STREETCAR MAKEOVER

While not built here, the PCC design was a result of research by the Detroit Edison Company's Chief of Research Dr. C. Floyd Hirshfeld. Privately funded by the Railway Presidents' Commission, this non-government project to redesign the surface streetcar was one of the earliest utilizations of what today is called the systems approach. The result of Dr. Hirshfeld's work was a vibration-free, silent-running, well-ventilated car modern in appearance. The first PCC's were produced in 1936, but Detroit didn't buy its first two until 1945.

From 1930 to 1937 and 1949 to 1962 Detroit used trackless electric trolleys, also known as trolley buses or trolley coaches. But the DSR never had much of a fleet compared to its conventional buses. In 1956, when streetcar service ended, there were 140 trolley buses versus 1,753 conventional buses. Trackless trolleys are rubber-tired bus-like vehicles powered by electric motors from a dual overhead wire system. Two wires were required since, unlike streetcars, trackless trolleys don't have the steel rail to provide a ground return-circuit.

The demise of all of these non-automotive systems — streetcars, trolley buses and interurbans — was due in large part to their inflexibility as well as perceived higher costs. Interurbans and — until the 1940s — streetcars had two-man crews, a motorman and a conductor to collect tickets.

Detroit was a city populated by those oriented to the automobile, who considered streetcars old-fashioned. Reflecting that public, Detroit's city fathers discarded the older public rail transportation concept, which today is being looked at again, nationally at least, as energy-efficient.

In 1995 Detroit again has light rail. It has been reborn twice since the last of the old streetcars ran in 1956. After a hiatus of 20 years, streetcars returned to downtown in 1976 with the opening of the City of Detroit's antique "Downtown Trolley" line running from Grand Circus Park down Washington Boulevard to Cobo Hall and later extended east along Jefferson to the Renaissance Center. This meter-gauge overhead electric trolley line primarily uses small, four-wheel vintage U.S.-built equipment made for Lisbon, Portugal, in the first decades of this century.

Detroit's latest rail adventure opened in 1987. It is the Detroit Transportation Corporation's "Detroit People Mover." This city-owned circular 2.9 mile elevated single-track line loops the city's downtown. Superficially a conventional system, it uses a standard gauge (4 foot 8.5 inch) track the same as most U.S. railroads and streetcar lines.

ADVANCED PROPULSION

It is in its mode of power — a linear induction system, considered space-age technology — that it's unusual. Most conventional rapid-transit street and subway cars still used in some North American cities have up to four DC traction motors. These are mounted on the truck frames and geared to the axles. In most cases 550- to 600-volt DC power is supplied from an overhead trolley wire or outside third rail.

In the People Mover's linear induction system there is no overhead wire or third rail. The cars lack conventional traction motors with their rotating armatures and stationary magnetic fields produced by coils, known as the stator. Rather, the People Mover has a broad 12-inch-wide flat conductor centered between the rails. This corresponds to the armature or rotor on an electric motor. The car's magnetic coils, which surround the center rail, constitute the stator.

When the center rail and car are energized, the car pulls itself along the center rail. Unlike the electric motor's moving armature the car itself moves. There are no moving parts other than the non-powered wheels. (The only other two systems like it anywhere in the world are in small transit feeder lines in Toronto and a 15-mile rapid-transit line in Vancouver.)

Until the 1930s, the regular steam railroads' frequent inter-city trains provided service to those wishing to commute between Detroit and its suburbs. Best remembered today was the Grand Trunk Western commuter service from Pontiac, Bloomfield Hills, Birmingham, Royal Oak and Ferndale to Brush Street Station (now the site of Renaissance Center), discontinued in 1983. Until the 1980s Amtrak provided commuter service from Ann Arbor on its regular trains. The New York Central ran commuter service until the 1920s from Grosse Ile to the Brush Street Station. Stops along the route included Woodward Avenue near today's New Center area. In the 1890s the New York Central also ran commuter service to East Jefferson Avenue near the Belle Isle Bridge on the industrial "Inner Belt" line.

Detroit's public rail history has been glorious. Today, with the exception of the People Mover and the Washington Boulevard trolley line, it is just a memory. Public bus transportation in Detroit and the suburbs hangs on, while public officials and citizens debate how to manage and finance it for the relative handful of customers who can't drive and must be served.

Stinson "Detroiter" monoplanes lined up at the Ford Air Port in Dearborn prior to delivery to Mexico for airline service, Sept. 8, 1928. (The Detroit News) **OPPOSITE:** *The original Wayne County airport terminal building was still serving airline passengers (and pre-jet age airliners) when this photo was taken June 27, 1962. The 1930-vintage building still stands at Detroit Metropolitan Airport's northeast corner in 1995, and may become a museum. (Detroit Free Press)*

The Sky's The Limit

Airfields open new vistas in transport, commerce

by Janet Braunstein

A s the 20th century approached, inventors like Wilbur and Orville Wright were still struggling to solve the puzzle of mechanical flight. They succeeded by 1903, but it wasn't until the First World War that pilots were trained in numbers, and the first Detroit area airfield was marked off by aviation enthusiast and real estate broker Henry B. Joy in 1917. Joy's 600 acres of marshland northeast of Mt. Clemens were leased to the U.S. government that year and activated as a military installation July 1, 1917.

The field was named in honor of 1st Lt. Thomas E. Selfridge, the first American to lose his life in an aircraft accident. The pilot gunnery school opened in April 1918, and the field's first fatality occurred on June 26, 1918, when 2nd Lt. John P. Boyle crashed into a pasture. Selfridge became the home of the 1st Pursuit Group, whose aces included Capt. Eddie Rickenbacker, who later founded Eastern Airlines. The government bought the field from Joy in 1921 for $190,000.

The Roaring Twenties brought an airport-building boom to Detroit and aviation to the civilian traveler. Henry Ford built Ford Air Port in 1925 on 260 acres on the Dearborn site now occupied by Ford's Dearborn Test Track on Oakwood. The same year, Ford bought Stout Metal Airplane Co. and began air freight service from Detroit to Chicago via the Ford Air Transport Service. In 1926, a separate company, Stout Air Service, began flying passengers between Ford Air Port and

Grand Rapids. The same year, the Grosse Ile-based Aircraft Development Corp. built an airship mast at Ford for its ZMC-2, the world's first — and only — metal-skinned zeppelin.

The mast was used only twice, once by the Army and once by the Navy, before it was torn down in 1946. By 1928, Ford Air Port had the world's first concrete runways. The paved runway is considered Henry Ford's greatest contribution to aviation.

NOISE COMPLAINTS

Aircraft Development Corp. set up its headquarters in 1926, after receiving a $300,000 Navy contract to build the ZMC-2, on the southern end of Grosse Ile on property where Ransom E. Olds built a summer home in the early 1900s. In 1927, the state of Michigan leased a five-acre strip of the property for use by the Naval Reserve Aviation unit formed at the Detroit Naval Armory in 1925. The move came after riverfront residents complained about the noise the pilots made flying a pair of seaplanes from the Detroit River to Selfridge and back.

The new seaplane base, which opened in 1929, had the world's first runway lights that were turned on automatically by weathervane control to outline the runway most directly into the wind.

Another air field was operated in the 1920s at 101/2 Mile and Gratiot by Packard Motor Car Co., which was building diesel aircraft engines and hired a team of test pilots to experiment with airplane designs. Earlier, Fisher Body flew "Jennies" from a grass field near the WWI production plant where it built them on Detroit's west side. When Stinson Aircraft startedbuilding cabin planes in Northville in 1926, it also utilized a nearby grass field.

The first landing at Detroit City Airport was made in 1927 by a Ford "Tin Goose" on the still unfinished airstrip. The airstrip was finished that year, and fenced and lighted during 1928 and 1929. By 1931, four airlines flew passengers from Detroit City to St. Louis, Toledo, Milwaukee and Chicago. But a natural gas tower that pilots used as a guidepost for finding the airport was deemed such a hazard that it, and the heavy

industrial and residential development around the airport doomed Detroit City to a short life as Detroit's major passenger terminal.

LAST STOP

Oakland-Pontiac Airport was opened in 1928 as Pontiac Municipal Airport and dedicated in 1929 as the final stop on the inaugural Michigan Air Tour. During the tour, sanctioned by a permit signed by Orville Wright, pilots in some 50 airplanes stopped at 30 air fields around the state.

There were three important fields in western Wayne County. The first was a square mile developed by the backers of Stinson Aircraft Co. for a new Stinson plant at Van Born Road and the Pere Marquette rail tracks, southwest of downtown Wayne. The concept for the aircraft plant, which operated from 1928 to 1948, was to deliver planes right from the factory door for customers to fly away. Since 1950 the site has been occupied by GM parts plants.

Ground for Wayne Field, now known as Detroit-Metropolitan Wayne County Airport, was broken in April 1929, but the first landing there wasn't until Feb. 11, 1930. The plane: a Stout Air Lines Ford Tri-Motor carrying local dignitaries from Ford Air Port in Dearborn. (In 1930, Stout Air Lines charged $18 to fly from Dearborn to Chicago, $10 to Cleveland and $5 to Toledo). By August. air mail service was shifted from Dearborn to Wayne Field. The two routes — Bay City-Detroit-Cleveland and Detroit-Chicago — were served by Thompson Aeronautical Corp., a predecessor of American Airlines, using single-engine Stinson Detroiters.

Willow Run, Detroit's most famous airfield, was a latecomer. Henry Ford bought the property in the 1930s to build Camp Willow Run, a place for sons of World War I veterans to live and work, growing and selling crops during the summer. In April 1941, after Ford Motor Co. won a federal contract to build B-24 bomber planes, Ford broke ground for the bomber plant and surrounding air field. By the end of 1941, the field had six paved runways, each with metal rods embedded in 14 inches of concrete to handle the heaviest four-engine aircraft in any weather. The bomber plant hit peak employment — 42,000 workers — in June 1943. By the end of the war, it had built 8,685 bombers and parts for a thousand more in less than four years.

World War II also led to the rapid development of the airport on Grosse Ile. In 1939, when the war broke out in Europe, the Navy was training just 23 pilots a month on Grosse Ile. Gearing up to train its own pilots and, urgently, many of Great Britain's, the Navy bought the reserve base property—including the landing field and huge balloon hangar of the failed Aircraft Development Corp., leased by the Navy since 1932 — from the state. In 1942, it nearly doubled the size of the base to 604 acres, paved a circular landing pad and three runways, added high intensity lights and installed equipment for instrument approaches. The base was designated U.S. Naval Air Station Grosse Ile on Dec. 9, 1942.

BUSH STINT

By the end of the war, more than 1,800 British cadets had received primary flight training on Grosse Ile. Among those who served at Grosse Ile was former President George Bush, stationed at the base in early spring 1945. After the war, Grosse Ile served as a reserve base until 1969, when the Navy closed it and moved the reservists to Selfridge. In late 1970s, the government turned over most of the property to the Township of Grosse Ile, which converted it to a 440-acre general-aviation airport and a 120 acre industrial park.

In the end, Wayne field became Detroit Metro, southeast Michigan's main passenger and freight airport, and Pontiac Municipal Airport became Oakland-Pontiac Airport, the region's busiest corporate and general aviation site.

Ford Air Port in Dearborn was shifted to use as an automotive test center during the 1930s. It was revived as an airport for military uses during WWII, but converted permanently to a vehicle proving ground in 1947. Dearborn Inn, arguably the world's first airport hotel, still stands, beautifully preserved and open for business, across the street from the proving ground. The last known use of the place as an airport occurred in 1975 when special permission was obtained for Northwest Airlines to land a DC-3 which it was donating to the Henry Ford Museum across Village Road from the Test Track.

Detroit City Airport was found inadequate for steady passenger service by large airplanes during the 1940s. In addition to the pesky natural gas tower,

which stood for at least two more decades, City Airport was bounded by two cemeteries and other dense development, all of which limited the direction from which planes could make instrument landings and limited ultimate extension of the runways.

Passenger traffic moved to Willow Run from Detroit City in 1946. Kaiser-Frazer bought the bomber plant to build cars (General Motors took over the plant from K-F in 1953). The University of Michigan bought the Willow Run Airport and Army Base from the government for $1 in 1947. The airport was state-of-the-art for the time, with hangars equipped with in-ground fuel-servicing pits and fast-acting bay doors; a complete lights system with emergency backup, a meteorology center and a control tower.

In 1947, a $1.5 million remodeling project converted a former B-24 hangar to a then-luxurious passenger terminal, and all seven airlines serving metro Detroit operated there.

OVERSHADOWED

But Willow Run was eventually eclipsed as well, by the jet age, which required even larger, more modern airfield and terminal facilities. In 1947, Wayne Field was renamed Wayne Major Airport. In 1954, Pan American World Airways gave Detroit its first transatlantic service, joined at Wayne by BOAC in 1954. In 1957, the airport was given its current name. In 1958, American and Allegheny airlines moved their scheduled passenger operations to a newly completed home at Detroit Metropolitan Wayne County Airport. They were followed by Northwest Orient and Delta in 1959, and by 1960 all seven of Metro's original airlines were offering jet service.

Ironically, for some 18 years after WWII, a debate over where to build a replacement for Willow Run simmered without solution. The aim was to find a location closer to the population, then clustered in and around Detroit proper. The choices included Windsor, Ontario, passengers could cross the river in five to 10 minutes, a great improvement over the 45 minutes it took to drive from downtown Detroit to Willow Run. During the debate, development continued at Metro Airport, once envisioned solely as a freight site, and the population, industry and business expanded north and west of Detroit. By the 1960s, Metro was equally convenient to most of the southeast Michigan population.

Typecast in Their Role

News media were not only the promoters of their day, but major players as well

by Janet Braunstein and Maureen McDonald

As 1895 dawned, the *Detroit Free Press* was being printed on machines just a generation or two advanced from medieval Gutenberg presses. News from other states and nations took days, weeks or even months to reach readers in Detroit.

A century later, newspapers still exist, but they can as easily be generated by a program on a readers home computer that automatically surfs a hundred on-line news offerings as by a printing press. News from around the planet is beamed by satellite in split seconds to television screens by a hundred or more cable channels. And computer software programs like those of magnate Bill Gates' Microsoft have made computers so easy to use that few offices — at home or at work — are without one.

It was only in 1892 that the *Free Press* had become the first newspaper in Michigan to install linotype machines, which used a keyboard at the machine to cast lines of type automatically from molten lead. In 1898, the paper replaced single-column headlines with multi-column headlines and used its first drawings in news columns. In January 1900, it used its first photo on the front page — a shot of Britain's Windsor boys, foretelling the continuing American fascination with England's royal family nearly a century later.

With no radio, television, cable or computers, newspapers were the only means of distributing information in the bustling industrial city of Detroit. "There

Detroit market by merging their business operations in a joint operating agreement that let the two papers — and their corporate owners, Knight-Ridder and Gannett — share production and delivery expenses and split profits.

The boom in the auto industry also brought the business press to Detroit, starting with the newsletter *Ward's Automotive Reports*. Starting in 1924, Alfred O. Ward paid people to count cars moving down the assembly lines at the factories of the city's dozens of auto manufacturers; the point of the exercise was to allow investors to tell whether the auto companies were solvent or not. *Ward's*, as it is known, remains the statistical bible of the auto industry; subscribers pay $1,000 year for the weekly tallies on auto sales, production, factory up- and down-time, and new models, technologies and factory sites. Sister publications include *Ward's AutoWorld, Ward's Automotive International, Ward's Engine & Vehicle Technology Update, Ward's Automotive Yearbook* — and a new monthly launched in late 1994, *Ward's Automotive International Focus on China*.

The weekly (once daily) *Automotive News*, read by some 80,000 car dealers, automaker executives, industry analysts and business reporters nationwide, was founded in 1925. Crain's Communications, a Chicago-based company, won the bidding when the widow of its founder put it up for sale. The publication was losing a half-million dollars a year when Keith Crain, son of Crain's founder J.D. Crain, moved his family to Detroit and took over. Crain found ways to cut costs, boost revenues and break even within six months. In 1977, Crain's bought *AutoWeek*, a magazine for car and racing enthusiasts, from a Reno, Nev., group and moved it to Detroit two years later. In January 1986, *AutoWeek* was converted from a grainy tabloid to a slick magazine. Meanwhile, Crain's Communications moved its corporate financial operation to Detroit in 1983 and its circulation department in 1985. The same year, the company added another publication, founding *Crain's Detroit Business*, sister to *Crain's Chicago Business*, then seven years old. In 1983, Crain's also bought *Monthly Detroit*, a city magazine, from its Cleveland owners and changed its name to *Detroit Monthly* in 1986. In 1995,

was no other form of news," says University of Michigan communications professor Jim Buckley. "Papers were divided by politics as well as class."

The *Detroit Times*, founded in 1900 by James Schemerhorn; the *Detroit Evening News*, founded in 1873 by James E. Scripps; and the *Detroit Journal*, founded in 1883, shared the busy market with the *Free Press*, founded as the *Democratic Free Press and Michigan Intelligencer* by John Pitts Sheldon in 1831. Papers such as the *Detroit Abend Post*, a German-language paper, served Detroit's many ethnic communities. During the 1870s, farm towns outside Detroit each had their own daily or weekly newspaper. Like the *Birmingham Eccentric*, founded in the 1870s, many turned into suburban papers after the interurban street cars connected previously isolated farm communities to the larger metropolitan area in the 1920s with the spread of auto factories to the Detroit outskirts.

WINNOWING DOWN

Throughout the 1800s and until purchased by Jack S. Knight in 1940, the *Free Press* was generally owned — either in part or in whole — by a succession of editors, generally its owner. Newsroom regimes changed hands each time the paper did; the era of corporate-think in Detroit daily newspaper journalism was yet to arrive. The *Detroit Evening News* remained independent until 1985 thanks in part to a lengthy legal battle James Scripps fought against his younger brother, E.W. Scripps, that ended when James traded his shares in the Scripps family papers owned in towns like Cleveland, St. Louis and Cincinnati for E.W.'s shares in the *News*. As Detroit grew, the number of players in the daily metropolitan newspaper market shrank. By the 1920s, the *News* had absorbed the *Detroit Journal*; in 1960, it bought the *Detroit Times*, which had been owned by the Hearst Corporation, one of the nation's largest chains. Thirty years later, the *Detroit Free Press* and the *Detroit News* ended their long battle over the

OPPOSITE: *When this 1948 Hudson pioneering television commercial was broadcast live on the Dumont Network, the nation's strongest TV market in New York City consisted of only 200,000 receivers. (Chrysler Historical Collection) In 1995, Detroit TV anchors Rich Fisher and Jerry Hodak present the news live several times daily over Channel 2. (WJBK-TV)*

a new publication was launched, *Crain's Nonprofit News.*

BEFORE BOOB TUBES

But back at the turn of the century, newspapers were king — at least for a while. For the most up-to-the-minute news, Detroiters would flock outside downtown newspaper offices to glean bits of information doled out by animated bulletin boards in front of the buildings. The closest thing to broadcasters were newspaper hawkers, who would shout two-word headlines to the throngs of pedestrians on the streets.

But on Aug. 20, 1920, a new era dawned. That evening, the nation's first radio station made its first broadcast from a studio in the corner of the *News'* city room. Over equipment ordered from inventor Dr. Lee DeForest, who cleared the way for commercial broadcasting by adding the radio tube oscillator to his earlier inventions of the radio tube triode and the radio amplifier, the *News* broadcast a phonograph recording of "Roses of Picardy." After receiving numerous responses from ham radio operators assuring that the song had come through just fine, the station played a recording of "Taps." Nightly music broadcasts continued for the rest of the month with no mention in the

pages of the *News* or any other Detroit paper. Then, on Aug. 31, the *News* ran an item announcing that it would broadcast results of state, congressional and county primary elections in southeast Michigan as they came in that night.

From then on, the station continued broadcasting news, and moved its studio to bigger facilities on the fourth floor of the *News* building on Lafayette and Second. The newspaper ran instructions telling readers how to construct their own crystal receiving sets inexpensively. On Sept. 6, 1920, the station broadcast news flashes on a boxing match in Benton Harbor between heavyweight champion Jack Dempsey and challenger Billy Misky. That, more than anything else, doomed the animated bulletin boards outside newspaper offices.

The government call letters eventually issued the *News'* radio station were first WBL, then WWJ. The first call letters issued to a commercial radio station were KDKA for the Westinghouse Electric and Manufacturing Company's station in East Pittsburgh, Pa., so it often turns up in history books as the nation's first radio station. But inventor DeForest himself helped verify that the *News'* station was the first; it broadcast for a full 11 weeks before KDKA came on the air.

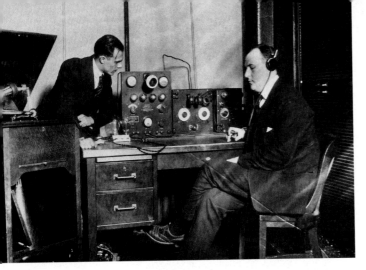

The nation's first radio news broadcast in August 1920 originates from a corner of the second floor newsroom at The Detroit News building on West Lafayette. (WWJ)

By 1922, WWJ's transmission range reached to 1,500 miles on a clear night; a U.S. Navy station at Bordeaux, France signaled that it had received WWJ loud and clear. The same year, Herbert Hoover made his first radio speech over WWJ from St. Paul's Episcopal Cathedral, where the station had set up equipment to broadcast the Easter Sunday service. Hoover, then U.S. Secretary of Commerce, decrying the then-stunning profits reaped by AT&T's early sponsored programs in 1923, said "It is inconceivable that we should allow so great a possibility for service to be drowned in advertising chatter."

Detroit was a marvelous market for the new medium. Its population burgeoned from 993,739 in 1920 to more than 1.5 million by 1926. The city boasted 200 movie houses (where the public could view motion picture news) and employed workers in some 2,000 different industries. And at a time when baseball truly was the nation's pastime, it was the home of the Detroit Tigers. In 1924, WWJ brought in a new broadcaster, Edwin L. (Ty) Tyson, to read news on the air. But Tyson was soon helping invent a new on-air job: sportscaster. Tyson covered powerboat races that, through the magic of the airwaves, made Detroit's Gar Wood a national motorboat hero. In the fall of 1924, Tyson made the Midwest's first radio broadcast of a football game, from the University of Michigan.

And on April 19, 1927, Tyson made the world's first broadcast of a major-league baseball game, from Tiger Stadium. Tyson's distinctive style became so popular with listeners that when the Tigers made it to the World Series, some 600,000 Detroit fans helped convince baseball officials that he should be allowed to broadcast the contest instead of some specially selected broadcaster not associated with either team.

In 1936, WWJ moved out of the *News* building into a new office across the street with five sound-proofed studios on three floors. In 1941, WWJ added a sister operation: the first FM station in Michigan, with studios at the top of the 47-story Penobscot Building, then Detroit's highest skyscraper.

VIDEO AGE DAWNS

The next leap in technology came just five years later. On Oct. 23, 1946, the *News* introduced Detroiters to television using closed-circuit monitors at the city's convention hall. The audience could watch the action on the stage or watch it appear simultaneously on screen. On March 4, 1947, WWJ began daily experimental TV broadcasts of test patterns and studio presentations. The date marks the founding of WWJ's TV station, which later became known by its current call letters, WDIV (Channel 4). There were fewer than 100 TV receivers in Detroit. Yet WWJ-TV was joined on the air on Oct. 9, 1948 by WXYZ-TV (Channel 7), and a few weeks later, on Oct. 24, by WJBK (Channel 2). Like WWJ, WXYZ was a radio station that added TV; during the 1930s and '40s, WXYZ radio brought the nation radio dramas including "The Lone Ranger," "The Green Hornet" and "The Challenge of the Yukon." (WXYZ radio later became WXYT.) WXYZ-TV made Soupy Sales a nationally famous comedian. By early 1949, Detroit was linked to the East Coast by coaxial cable; network television had come to Michigan.

For more than a decade, television was no threat to newspapers or even radio because it lacked the depth of the first and the instancy of the second. "Television was strictly a radio with pictures on it and they didn't do it very well," said longtime Detroit Channel 2 newsman Joe Weaver, who worked in radio from 1948 to 1960.

Once film was shot for TV, it had to be developed — a process that could take up to an hour — and then laboriously edited by cutting and splicing. But in the early 1960s a new invention — videotape — gave television the immediacy that earned it a distinct place alongside print and radio. Videotape offered much greater clarity than grainy film and could be popped out of the camera and shown on the air without developing. Instead of cutting and splicing, editors transferred images from one tape to another, rearranging them as needed quickly and easily. "Videotape really took television out of left field," Weaver said.

By the 1970s, a new technology began offering metro Detroiters, at least those in the suburbs, a choice between receiving their TV signals over antennas — a few strong VHF channels and a few very weak UHF channels — or via cable, which offered clear reception on nearly every spot on the dial. Grosse Pointe was one of the nation's first communities wired for cable;

Detroit proper, one of the last. Barden Cablevision, already serving Inkster, Romulus and Van Buren, turned in its bid for the city of Detroit franchise on Dec.10, 1983. The job of laying some 2,100 miles of cable in Detroit began January 1986 and was completed in 1992. With cable, viewers who once had a choice of at most 12 channels became inundated with offerings, not all of them good, as attested to in Bruce Springsteen's song "57 Channels and Nothing On." But in 1994, Detroit was the nation's eighth-largest cable market and, with 63 percent of households receiving cable, had a higher percentage of cable penetration than Los Angeles or Chicago, both larger markets.

During the 1980s, another new medium was born: computer on-line information services such as CompuServe, Prodigy and America Online. This time, the first Detroit newspaper to take advantage of the new technology wasn't the *News*. The *Free Press* began offering a full-service computer bulletin board over CompuServe in January 1993. CompuServe members nationwide can log onto the Detroit Free Press forum and exchange messages, read current newspaper stories, dig into the newspaper's past issues in an on-line library and communicate with reporters and editors. Periodically, like other forums, it holds live on-line conferences. The Free Press forum is part of Free Press Plus, an ongoing effort by the newspaper to use new technology such as facsimile transmission (more familiarly, "fax") and computers to win itself a continuing place in readers' lives.

Certainly, that aim is deserving of plenty of study. Media visionaries predict that within a few years, cable, telephone, data and broadcast services may all be combined so that a user can punch up any or all on one computer monitor. Already, there is software that enables users to create their own personal newspapers by letting them make a series of choices about what kind of news

they want and how they want it to look. The program goes on-line automatically to locate the news each user wants — from one source or from the hundreds of magazines, newspapers, wire services and other news services on-line — and then prints out something that looks just like the package that hits our doorstep every morning.

Health Care 'Reform'

Rise of modern facilities and practices gave sick, injured new hope

by Martha Hindes

TOP: Rendering of the Veterans Administration Medical Center, nearing completion in 1995 in Detroit's University Cultural Center adjacent to the Wayne State University Medical School and just south of the Rackham Building, home of ESD. (SH&G)

While "hospital" and "deathbed" were hardly synonymous a hundred years ago, their relationship was very close. Those who couldn't be cared for at home and were hospitalized probably were very poor or gravely ill. Horse-drawn ambulances that brought in the sick also took bodies away.

Detroit's earliest hospitals, begun around the Civil War period, were housed in their own specially-built structures by the end of the 19th century. An early St. Mary's Hospital downtown was operated by its founding Daughters of Charity Catholic sisters who later established Providence Hospital, now in Southfield.

The Protestant-founded Grace and large Harper hospitals handled many of the city's less fortunate, while Woman's Hospital and Foundlings' Home, later called Hutzel, was a haven for unwed mothers and abandoned children. Those hospitals became a core of the Detroit Medical Center about a decade ago.

In the early 1900s, when city officials were unsuccessful in erecting a new hospital to treat the burgeoning Detroit population, auto pioneer Henry Ford took over the job. Ford, who had brought droves of workers to his auto plants, got the Henry Ford Hospital built just in time for World War I and the deadly 1918 influenza epidemic.

In sparsely settled Livonia 35 years later following a nearby General Motors plant fire, Catholic Felician

sisters answered urgent pleas by building another St. Mary's Hospital.

While Detroit had about 10 hospitals at the turn of the century, their number — spurred by medical advancements and enterprising physicians — increased to more than 100 around the 1930s and 1940s.

With segregaton still a daily reality, a group of black doctors in May of 1918 founded Dunbar Hospital on Frederick Street near the present day medical center so their patients could get the same care whites had been getting for years.

The facility, named for nationally known black poet Paul Laurence Dunbar, later branched out as Park-side hospital on the city's northwest side, then permanently closed in 1962. The original structure now is a Detroit historic site.

As the century progressed, the hospital count began to decline through a trend toward mergers and acquisitions. According to the Southeast Michigan Hospital Council (SMHC), the seven-county metropolitan Detroit area had 73 separate hospital complexes in 1983, and has 57 today.

The transition in health care during the past century has been a journey from easing suffering to maintaining health. And hospitals — in the forefront of the evolution — have been restructured in the process.

Staying healthy in the 1890s was a daunting task, according to Prof. Nicholas Steneck, director of the Historical Center for Health Sciences at the University of Michigan.

Childbirth was hazardous, and childhood illnesses and communicable diseases took their toll. A century ago, the average lifespan for men was 47 years while the average woman lived to be only 46.

Americans in 1895 were vulnerable to a long list of deadly illnesses, with heart disease and influenza topping the list. But pneumonia, tuberculosis, diphtheria, typhoid, malaria, measles and whooping cough marched in deadly step behind those two killers. And polio and scarlet fever were common in Michigan at the time, along with the cancers and cardiovascular illnesses still prevalent today.

Many vaccines had not yet been developed and the first antibiotics would not be discovered for another 30 to 40 years.

Surgery during the late 1800s was still a perilous adventure, conducted in hospital amphitheaters on patients drugged with gases such as ether. Those who survived were treated by nurses working 24-hour shifts. But changes were on the way.

"A hundred years ago, we were right at the begin-ning of the emergence of the modern era of medicine," said Steneck.

Doctors of 1895, armed with the germ theory of illness discovered a decade before, were learning to operate under sterile conditions. The terrible 90 percent death rate for surgery patients was finally reversing itself.

Diagnoses would grow more accurate as turn-of-the century hospitals began to acquire the forerunners of modern equipment just being developed, such as elementary x-ray machines.

Today, southeast Michigan hospitals have become highly technical, specialized centers with satellites for outpatient surgery, walk-in emergencies, and mental and physical health maintenance. The Detroit Medical Center sends trauma cases to its Receiving facility, and children to Children's Hospital. St. John Hospital and Medical Center, on Detroit's east side, does more open heart surgeries than any other hospital in Michigan.

Building expansions, such as a seven-year project nearly complete at Royal Oak's 40-year-old Beaumont Hospital, have added the latest specialized technology while improving surroundings for patients and families.

And smaller independents (the old Detroit Osteopathic Hospital, now a Horizon Health System member, is one) have joined as partners in widespread health care groups pooling resources, patient populations and area medical school affiliations. Some, such as Henry Ford Health System, have health maintenance organizations, or HMOs, to provide cradle-to-grave care through physician and satellite networks.

Change has been spurred by previously unimaginable advances in medical care that have helped increase the likely longevity for a baby born today. As of 1993, the Michigan Department of Public Health estimates a newborn white female will live 79.1 years, a white male 73.4 years. Blacks haven't gained as much, with current life expectancies of 72.9 years for females and 63.8 years for males born today.

Other factors shaping the state's health care facilities are stringent cost containment and a state ban on more inpatient beds. Hospitals have adapted by building care centers in outlying areas where people were moving. But many surgeries that once meant weeks of hospital bed recovery can be done at outpatient clinics with patients sent home to rest.

"Now they can literally make the blind see, the deaf hear and the lame walk," says Donald P. Potter, president of the SMHC.

"Hospitals are no longer the alms houses where people go to die."

IV.

0

Science's Scions

Enterprises ripen as fruits of invention
— inevitably, some of them wither, too

We've Been Helping Detroit Industry Communicate Since The Information Highway Was Just A Dirt Road.

It was back in the late 60s that two very important events occurred. One was the creation of ARPAnet, the progenitor to today's hottest communications medium – the Internet. The other was the creation of French & Rogers, today one of the Midwest's leading business-to-business advertising and marketing agencies.

We Happen To Think Technology Is Exciting.

We've enjoyed many successful client relationships over the past 27 years because business-to-business marketing is our business. Unlike some agencies, we're just as comfortable digging into a new manufacturing process or engineered product as we are developing a slogan for a new brand of toothpaste.

We Already Know Your Market, Your Customers And The Best Ways To Reach Them.

Our agency is populated by talented people with technical experience. Some have engineering degrees and others are former editors. We blend firsthand knowledge and expertise in a variety of industries with outstanding creative abilities, to find a unique twist that can set your product or service ahead of your competition.

We Provide The Full Range Of Marketing And Communications Services.

We've produced award-winning work across the board – advertising, public relations, literature, direct marketing, packaging, trade shows, videos and interactive media – and we're adding Internet marketing to our list. We also work as a marketing resource for our clients, helping to analyze markets, establish goals, develop strategies, and implement programs.

Let us help you get your share of the next Technology Century. Contact Jim Meloche at French & Rogers today.

French & Rogers
5750 New King Street, Suite 330
Troy, MI 48098-2696
TEL: (810) 641-0044 • FAX: (810) 641-1718
E-MAIL: 76247.2752@compuserve.com

Salt of The Earth

Vast underground deposits made Detroit chemical industry center

by Charlotte W. Craig

TOP: *International Salt's mine beneath Detroit was so huge a miniature city was created underground. Here at 1,135 feet beneath the surface in 1962, Jeep trucks brought down in parts were reassembled in this "machine shed" to be used for transportation. (Detroit Free Press)*

Before Henry Ford's first car ever rolled down a Detroit street, there was the salt of the earth ... and a different man named Ford.

He was Captain John B. Ford — a remarkable personality who made and lost several fortunes during his life, earned a name as the father of America's plate glass industry and founded Detroit's first chemical company at the age of 79.

Even the earliest settlers in Detroit were aware that the area was underlain by a wide layer of rock salt — discovered, no doubt, when they attempted to sink wells for drinking water. A few enterprising folks began using and selling salt brine, obtained by pumping salt-water wells or by flushing holes dug into the salt layer. The brine was useful for tanning and other purposes.

But it was not until 1890, when Captain Ford founded the Michigan Alkali Co., that Detroit's great salt deposits began to be exploited on a commercial scale.

Ford, a self-educated, energetic and independent man, had tried an arm-long list of careers by the time he came to the Detroit area. He had been, successively, a harness- and saddle-maker, a dry-goods-store owner, flour-mill operator, kitchen-cabinet manufacturer, iron rolling-mill and foundry owner, shipbuilder and glass-maker, all in the state of Indiana. It was his business of

169

building river boats during the Civil War that attached the title Captain to his name for the rest of his life.

At almost 70 years of age, John B. Ford moved to Pennsylvania and began the Pittsburgh Plate Glass Co. It was the appetite of this fast growing business that eventually led Ford to the banks of the Detroit River and the little town of Wyandotte, where he shrewdly saw a vast source of salt-based soda ash, a primary ingredient for glass making.

RELINQUISHED REINS

Ford was 79 when he founded Michigan Alkali. Following an eye operation, he officially retired from business in 1897, turning most of the operations of his various firms over to his sons and close associates. But in 1898, he gave his blessing to creation of the J.B. Ford Co. to package and market some of the alkali-based consumer byproducts coming out of Michigan Alkali.

In "Landmarks of Detroit," published in 1898, authors Robert Ross and George Catlin reported proudly that "Some $7,000,000 of capital have become engaged in the manufacture of salt and salt products, none of it being invested actually within the city, but all in its immediate vicinity. These works now employ in the neighborhood of 2,000 hands ..."

Captain Ford died May 1, 1903, at the age of 91. But the story of his family and his many businesses was far from over.

The Captain's offspring and descendants, known locally as "the chemical Fords" to differentiate them from "the automotive Fords," went on to become powerful and active in a long list of civic and commercial enterprises, including Libbey-Owens-Ford, Huron Portland Cement, Parke Davis & Co., Manufacturers National Bank, National Bank of Detroit, Wyandotte Transportation Co., and the Detroit Symphony Orchestra.

On Jan. 1, 1943, Michigan Alkali — with executive offices in the Ford Building in downtown Detroit — and the J.B. Ford companies officially combined to become Wyandotte Chemicals Corp. The consumer products marketed by the J.B. Ford Co. had been primarily cleaning agents bearing the Wyandotte brand name. A turn-of-the-century advertisement published in Cleveland advised, "Any man who gets his wife a sack of Wyandotte Cleaner and Cleanser is saving her strength and her time."

After the merger and formation of Wyandotte Chemicals, the cleaning business continued to grow. A 1956 Detroit Free Press article described Wyandotte as No. 2 in overall chemical production after Dow Chem-

ical in Midland, but No. 1 in the world in production of specialized cleaning compounds.

Wyandotte Chemicals remained a privately-owned corporation until 1956, when 100,000 shares of stock were offered for public sale. At that time, the company operated three plants in the downriver Detroit area, the world's largest limestone quarries in Alpena and four other plants scattered across the country. The company's dry ice works was the largest in the world. Principal products included soda ash, bicarbonate of soda, caustic soda, calcium carbonate, calcium chloride, chlorine, dry ice, synthetic detergents, "wetting agents," glycols, ethylene and propylene dichloride.

At the time Wyandotte Chemicals "went public," the U.S. Commerce Department reported 239 chemical companies in the Detroit area, employing nearly 18,000 people. Among them were plants operated by Allied Chemical, DuPont, Monsanto, Pennsylvania Salt Manufacturing Co. and Sharples Chemicals.

STOCK SLIDE

In 1969, just 13 years after Wyandotte Chemicals issued its first stock, the West German chemical conglomerate BASF (Badische Anilin & Sodafabrik) bought a 14-percent interest in the company. Faced with growing environmental concerns and costs, Wyandotte's earnings had begun to slip; 1968 earnings were 65 cents a share, versus $1.43 just a year earlier. By November of 1969, BASF had bought 98.5 percent of Wyandotte Chemicals for a total of about $95 million.

Wyandotte Chemicals officially became BASF Wyandotte in 1971. The move was symbolic of the international nature of Detroit's chemical industry, which from the beginning has formed a strong link with world markets.

In 1978, the company announced it would discontinue production of synthetic soda ash in Wyandotte, laying off about 450 hourly workers and relocating about 250 salaried employees. About 1,900 employees remained at the plant after the layoffs.

Today, there are many changes in the company Captain John Ford envisioned and built. For one thing, the name Wyandotte dropped away in 1985, leaving BASF Corp. as the name for all the Detroit-area facilities. Yet, reduction and consolidation have not been the only name of the game. The number of Detroit-area employees is higher; products are more diversified.

True, some of the original Wyandotte Chemicals buildings are gone now. But BASF has invested more than $130 million in new buildings in the Wyandotte complex. At the south end of town, the company paid more millions to tear down 84 acres of old chemical buildings and have the land tested and cleaned for use as a park.

There's an automotive refinish coatings office and research center in Southfield, an automotive original equipment coatings office in Troy, two auto finish production facilities in Detroit and a refinish coatings facility in Dearborn. In Livonia, there's a BASF office and plant for making urethane specialty products used in applications as diverse as bowling balls and shoe soles. There are about 2,300 BASF employees in the Detroit area.

Hard on the heels of Captain Ford's Michigan Alkali Co. came a Detroit plant built by the giant international Solvay company based in Belgium — drawn to the banks of the Detroit River by the vast salt deposits.

Philanthropist

Ernest Solvay, the founder of the company, was known as the Belgian Carnegie and launched almost as many social as scientific experiments. Ernest Solvay first patented his process for making soda ash in 1861. The Solvay process involved dissolving salt in ammonia water, through which carbonic acid gas was passed. The precipitated alkali was furnaced as bicarbonate of soda, yielding carbonate of soda or soda ash.

The first Solvay facility in the United States was established near Syracuse, N.Y., in 1881 and became operational in 1884. A Detroit plant was established not too long after — probably in the mid-1890s — and was known as the Solvay Process Co.

Solvay and his company appear to have been rather progressive for their time. A book about the man and the company, "Solvay, A Giant," by Maxime Rapaille (published 1989) says the company had established a general policy throughout its facilities adopting an eight-hour work day by 1907. Sickness benefits, too, were adopted by the company in the late 1800s.

Solvay Process, located in the Delray neighborhood of Detroit on West Jefferson, also had salt mining properties on Zug Island in the Detroit River. In the 1920s, Solvay Process was one of several companies that com-

Solvay Automotive supplies blow-moulded, high-density polyethylene (HDPE) fuel tanks for vehicles such as this 1995 Oldsmobile Aurora. (Eisbrenner Public Relations/Solvay Automotive)

bined to form Allied Chemical, although the Solvay trademark continued. The company suffered a series of strikes during the late 1940s and early 1950s; the period also was financially difficult due to the fact that business was based primarily on one product.

It was at this point that plastic began to emerge as a product line for Solvay. It was a non-petroleum plastic, using chlorine, which already was made by Solvay, and acetylene, obtained from calcium carbide via the coal byproduct industry. A process using petroleum ethylene was later incorporated into operations.

Solvay owned some stock and a board seat within Allied Chemical (now AlliedSignal Inc.) until 1974. That year, Solvay bought the polyethylene production facilities of Celanese Corp. in Texas, near Houston. Because that purchase put Solvay in direct competition with Allied, Solvay sold its share in Allied and pursued its separate course.

In the 1980s, Solvay's U.S. executives recognized the auto industry as a major future market. Solvay Automotive Inc. was formed in 1990 through the merger of Hedwin Automotive Division, which already was part of Solvay, and Kuhlman Plastics Division, purchased by Solvay in 1990.

Fuel Lines

Corporate offices for Solvay Automotive are on West Maple Road in Troy; the old Detroit plant is long gone. The company today is the largest producer in the U.S. of polymeric fuel tanks; it also makes fuel systems, including pumps, level sensors, filler pipes, fuel lines, valving and safety devices, as well as other automotive blow-molded products such as seat backs, fluid bottles and air ducts. Non-automotive products include marine fuel tanks, large water filters, stadium seats and golf cart bumpers.

A separate company, D&S Plastics, a joint venture in which Solvay America is a part owner, has a technical center in Troy and a plant in Texas.

In 1994, Solvay Automotive had sales of $150 million and about 70 employees in the Detroit area. U.S. production operations are in South Bend, Ind.

Meanwhile, AlliedSignal Inc. has established its own presence in the Detroit area, with AlliedSignal Automotive World Headquarters located in Southfield and other AlliedSignal offices and plants situated in Detroit, Livonia, St. Clair Shores, Sterling Heights, Troy and Warren.

Counting plants in Boyne City and St. Joseph, AlliedSignal has about 3,000 employees in Michigan and a payroll of more than $111 million annually. The company operates two automotive engineering and research/development facilities and its four manufacturing plants make products as diverse as creosote, roofing pitch, gyroscopes, disc brakes and hydraulic booster power units.

While Captain John B. Ford and the Solvay company had pioneered the use of brine wells in the manufacture of alkali and related products, actual mining of rock salt from the beds beneath Detroit was begun in 1904 near River Rouge by the Detroit Rock Salt Co. The company's first deep shaft was started in 1906 and took four years and the lives of eight men to complete, according to Detroit Free Press clippings.

In 1912, the company was acquired by International Salt Co., although it retained the Detroit Rock Salt name for many years. About 1,000 acres eventually were mined by International Salt Co. under Detroit, Melvindale, Dearborn and Allen Park. Headquarters for the operations were on Sanders near Fort Street in Detroit. The mine employed 120 to 150 miners and support personnel through the 1960s and into the early 1980s.

A quarter of a mile beneath the surface, the mine was like a small, strange and sparsely populated city, with almost 100 miles of roadway formed by a complex of dark tunnels with white walls and ceilings. The tunnels were about 22 feet high and wound around huge pillars of salt left standing for extra support. Temperature in the mine was a more-or-less constant 57 degrees.

The Detroit-area salt mines have been many things to many people through the years — a source of work and income, a novel and spine tingling tour site, a potential air conditioning system, a bomb shelter and a space-age test laboratory.

At one point in 1950, Detroit Mayor Albert Cobo instructed Civil Defense coordinators to investigate the possibility of using the mine's underground corridors as a bomb shelter for most of the area's citizens, in the event of nuclear attack.

SALT SENSE

The notion was abandoned after it was pointed out that people would have to be taken down 10 at a time in mine elevators, and survival would depend on surface equipment.

A 1953 article in the Free Press said International Salt Co. had signed a contract with Wayne County permitting the firm to mine salt under Detroit-Wayne Major Airport (which became Detroit Metropolitan Airport in 1958) for the next century. Total value of the contract was estimated at the time to be almost $2 million — much of it based on a future per-ton royalty. Part of the plan called for bringing air up from the mine shafts to air condition the airport for free. There is no evidence, however, that the plan ever came to fruition.

In 1967, an old unused portion of the mine was leased by engineers of the Missiles & Space Division of LTV Aerospace Corp. as a test lab for inertial guidance systems. They needed a place free from vibration and noise; they found it in the dark beneath Detroit.

In 1970, International Salt was taking about a million tons of salt a year out of the Detroit mine, most of it sold for ice control on streets and highways; the remainder went toward making chlorine and water softeners and leather tanning products.

That same year, International Salt merged with American Enka Corp., which at the same time changed its name to American Akzo Corp., reflecting the majority interest held by the Dutch company Akzo N.V. — more world ties for Detroit's chemical industry.

International Salt closed its Detroit mine in January of 1983, the victim of a warm winter and heavy competition from cheaper Canadian rock salt. Millions of dollars worth of heavy mining equipment was disassembled and hauled out the way it went in — through the 5-by-6-foot elevator shafts. Many worn-out pieces simply were left behind.

The abandoned mine was bought in 1985 by Crystal Mines Inc., which ran tours through the site and laid plans for using it as a toxic and hazardous waste dump. However, public outcry against the plan was considerable, and formal permission to use the site as a dump never was sought from the Michigan Department of Natural Resources.

Tours through the mine were discontinued in 1986 and it now stands unused. However, International Salt — since 1992 renamed Akzo Nobel Salt, based in Pennsylvania — still maintains Michigan operations in Manistee and St. Clair.

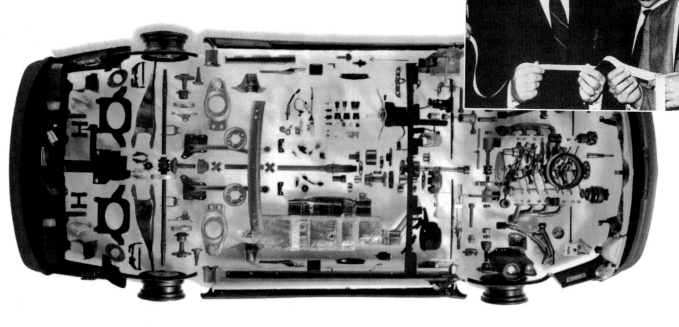

This plan view of an automobile illustrates the many parts Masco supplies to the auto industry. *(Masco)* **RIGHT:** *Richard Manoogian (left) and his father Alex, founder of the company, in a stock exchange photo in the 1960s. (Masco)*

Value in Diversity

Alex Manoogian's corporate descendant is worth study, root & branch

The Masco story began in 1920 when 19-year-old Alex Manoogian, the son of an Armenian grain wholesaler, emigrated to this country from his birthplace in Smyrna, Turkey, to escape the brutality that existed in that country following World War I. He arrived in New York with two suitcases, $50 in cash and a name tag around his neck..

He worked in a variety of jobs on the East Coast and moved to Detroit in 1924 where he learned to operate a screw machine. Manoogian wanted to capitalize on the higher wages being paid by the automotive industry, and he believed it would be easier to bring the rest of his family to America through Canada than through Ellis Island.

In 1929, an economically disastrous year, Manoogian and two partners began Masco Screw Products Company. The name was formed from the first three letters of the last initials of the partners with the addition of "co" for company. Before the year had ended, Manoogian had bought out his partners.

The young company faced its first crisis soon after it opened when oil from a faulty screw machine hose leaked through the floor, ruining thousands of dollars worth of new upholstered furniture on the floor below. The furniture manufacturer could have sued, but declined, probably because he knew that Manoogian had no money to pay for the damage. Instead, he agreed to extended payments. Manoogian stood by his promise and the damages were paid in full.

This old-fashioned screw machine was the basis for today's Masco Industries, founded in 1929 as Masco Screw Products Co. (Masco)

Masco's first major contract, a machining job for $7,000, came from the Hudson Motor Car Company, and Masco went on to become a supplier to many other automakers. The Big Three and their competitors were important customers, as were automotive suppliers such as Budd and Spicer. The company reported gross sales of $66,000 for the first full year of operation, and within two years was operating at full capacity.

In the meantime, Manoogian had saved enough to bring his family to America in 1931 and married Marie Tatian the same year.

Masco's sales exceeded $100,000 for the first time in 1935. A public offering of stock — the first of many to follow — was undertaken in early 1937 when shares were listed on the Detroit Stock Exchange. Some 120,000 shares were offered at $1 each.

NEARLY A PIPE DREAM

Manoogian's dreams almost went up in smoke in December 1937 when the plant caught fire. When notified, Manoogian raced to the scene and wept as he watched flames licking at the windows. But fortune smiled on the fledgling company when fire damage was minimized by the snow that had accumulated on the roof; it melted and helped douse the fire. Manoogian worked day and night to rewire his machinery and Masco was back in business within three months. However, it became the only money-losing year in the company's history.

In 1941, upon America's entry into World War II, the Detroit automotive industry and its component suppliers quickly converted to the manufacture of vehicles, aircraft and other vital products for the country's defense. The company's performance during the war led to substantial increases in sales and cash flow. Masco annual sales in 1942 crossed the one million dollar mark.

During the Korean Conflict, Masco returned to war production again, using the company's metalworking expertise in a variety of applications, including the development of a new artillery shell timing mechanism. This experience in precision-machined parts provided a platform from which to expand into newer, more complex metalworking applications.

Because of Masco's success with these more sophisticated parts, Chrysler Corporation asked Masco to produce a transmission shaft using a new process called cold extrusion. From this came other opportunities for expansion in the metalworking industry.

A major turning point in the company's history occurred in 1954 when Masco began manufacturing a few machined parts for a product that would revolutionize the plumbing market — the single-handed faucet. Initially the public snapped up the new product, but sales fell subsequently when a reputation for poor quality got around. Alex Manoogian applied his metalworking expertise to redesigning the product. Soon he acquired all manufacturing rights and the resulting "Delta faucet" launched the company into the building products industry.

DELTA DIVIDEND

By 1958 Delta faucet sales had passed the $1 million mark, bringing total Masco sales to almost $4 million. A year later, to organize the company's diverse businesses more effectively and meet the growing demand for the faucet, Masco began construction of a new plant in Indiana devoted exclusively to Delta production.

The same year, Manoogian's son Richard joined the company upon graduation from Yale, initially to help with the startup of the new faucet operation. Subsequently Richard utilized the cash flow being generated by faucet sales to embark on a friendly acquisition program. This paid off handsomely for Masco investors. An investment of $10,000 in Masco stock in 1958, when the expansion program was launched, would be worth some $30 million 35 years later.

Masco's acquisition program was based on corporate objectives that included:

- proprietary leadership positions;
- diversification of the product base to avoid dependence on any single market;
- leveraging the company's existing product development, manufacturing and marketing skills; and
- manufacturing only products that offered an above-average rate of return.

In 1960, Masco licensed Emco Limited, a Canadian company, to market Delta faucets in that country. Emco distributes certain Masco products in Canada and a number of Emco products are marketed in the United States in conjunction with Masco companies.

In 1961, Peerless Industries, Inc., became the first outside company to join the Masco family. The Peerless acquisition enabled the company to offer customers a broader line of plumbing products. That year also marked a name change from Masco Screw Products to Masco Corporation. The move reflected the company's diversification into the building products industry while preserving a name known in the automotive industry as a leading component supplier.

Diversification continued throughout the next several years as Masco acquired a number of complementary manufacturers of industrial and building products.

In a dramatic strategic initiative after 30 years of consecutive growth in sales and earnings, Masco Corporation in 1984 spun off Masco Industries as a new public company composed of its successful industrial component businesses, with approximately $400 million of net sales. The parent Masco Corporation became focused on consumer brand-name products for the home and family. Management believed each could grow faster as a separate entity by building upon its unique strengths in its specific market niches.

Masco Industries invested more than $1.2 billion in acquisitions and capital expenditures to enhance its positions in the industrial market it serves. The addition of new metalworking skills and technologies enabled the company to build upon its ability to engineer, manufacture and market high quality industrial products at competitive prices.

NYSE LISTED

In 1993, Masco Industries changed its name to MascoTech, Inc., and listed its shares on the New York Stock Exchange. The new identity reflected increased emphasis on technology and continued commitment to Masco traditions of supplying the automotive industry.

Masco Corporation, with sales of approximately $600 million in 1984 after the Masco Industries separation, grew dramatically over the next decade through a combination of internal growth and acquisitions. In 1985, Masco expanded its presence in the building products industry by merging with the leading cabinet maker, Merillat Industries. In subsequent years other cabinet manufacturers including Kraftmaid, StarMark and Fieldstone joined the Masco family, making the company the nation's largest kitchen and bath cabinet manufacturer.

Masco entered the furniture business in dramatic fashion in 1986 with the acquisition of two leadership companies, Henredon Furniture and Drexel Heritage. This established Masco as the largest manufacturer of high-end wood and upholstered furniture in the U.S.

A year later, Masco further expanded in the furniture business by acquiring Lexington Furniture Industries, the largest domestic manufacturer of mid-priced solid wood dining and bedroom furniture.

In 1988, a third member of the Masco family, Tri-Mas Corporation, was established as a separate public entity when it acquired nine industrial businesses from Masco Industries. TriMas' market positions include proprietary products for commercial, industrial and consumer markets. The company was listed on the New York Stock Exchange in 1991.

In the meantime, Masco Corporation completed its largest single acquisition in 1989 and rounded out its "good, better, best" furniture offerings when Universal Furniture Limited, a popular-priced firm, became part of the Masco family of companies. The move established a Far East manufacturing base and participation in international markets, as well as offering additional product opportunities such as wood flooring and motion upholstery.

The Masco family of companies today employs more than 64,000 people and has 270 manufacturing facilities with over 43 million square feet of manufacturing space in 22 states and 17 countries. Masco Corporation had 1994 sales in excess of $4 billion in the building and furnishings markets.

MascoTech, Inc., with 1994 sales of nearly $2 billion, is a leading supplier to the transportation industry with components, assemblies and subassemblies used in every vehicle manufactured in North America.

TriMas Corporation reported 1994 sales of $535 million as a manufacturer of fasteners, towing systems, specialty containers, precision tools and other products for commercial and industrial applications.

The Manoogian family and Masco have indeed come a long way from Ellis Island and the struggling little screw machine company of the 1920s.

A float glass line at a Guardian Industries plant. (Guardian Industries)

Clear Winners

Glass makers' fortunes rose with building, auto industries

Along with Ford Motor Company's Glass Division — established by Henry Ford at the Highland Park plant in 1921 as part of his effort to control production of all automotive components — Guardian Industries Corp. is one of two Detroit-area companies that went into glass production to serve automotive needs and later expanded to architectural glass. Guardian also exemplifies a successful Detroit-based technology company which has expanded into other businesses on a worldwide basis.

The history of Guardian dates to 1932 when Guardian Glass Company was formed in Detroit to fabricate glass for the automotive industry. William Davidson, president and chief executive officer, assumed control of Guardian Glass Company in 1957.

Through his leadership, the company expanded its business significantly and was producing windshields for new car and truck manufacturers as well as being the world's largest school bus window supplier by the mid-1960s.

Guardian Industries Corp. was incorporated in 1968, becoming the successor to Guardian Glass Company. From 1968 to 1984, Guardian was a publicly-held, New York Stock Exchange-listed company. Early in 1985, Davidson took Guardian private in a transaction involving the purchase of all shares of the company not previously owned by him.

In 1968, Guardian saw the need and opportunity to become a raw float glass manufacturer in addition to a glass fabricator. By this time, the company was a major user of raw glass, but was obliged to obtain this glass from its competitors. Yet, building a modern float glass plant at the time required an enormous amount of capital in relation to Guardian's sales and equity then. The money was raised and in 1970, Guardian built its first float glass plant in Carleton, Michigan, becoming the first new company to enter the North American glass manufacturing business in fifty years.

We've Shared Your Vision Through Thick And Fin.

Ever since cars first hit the road there's been a drive to improve and innovate. And from the start PPG has worked with designers and engineers to help bring their ideas to life. PPG products like laminated window glass, curved windshields, windshields with integrated radio antennas, solar control and privacy glass have helped revolutionize the aesthetics, comfort and safety of today's automobiles.

At PPG, we're proud of the part we've played in the growth of this industry and we're looking forward to creating tomorrow's visions.

PPG Industries, Inc.   Automotive Technical Center   5875 New King Court   P.O. Box 3510   Troy, Michigan 48007-3510

Since opening the Carleton plant, Guardian has added five new domestic float lines, including another in Michigan. As part of the company's global vision, Guardian expanded its operations into international markets. In 1981, Guardian entered the Western European float glass market through the construction of a new plant in Luxembourg. Guardian has subsequently added two new plants in Spain and a second plant in Luxembourg. In 1991, Guardian led investors into Eastern Europe by constructing a new plant in Oroshaza, Hungary, making it one of the largest Western investments in that country.

Guardian has also continued to expand in other parts of the world. In 1990, a joint-venture float glass plant was completed in Venezuela, one of only four float glass plants in South America. Concurrently, Guardian embarked on a strategy to penetrate the Asian glass market, through the construction of float glass plants in Thailand and India. These plants began operations in 1992 and 1993, respectively. To further its international competitiveness, Guardian has recently begun construction at two new float glass plants in the U.S. and Saudi Arabia.

COMPANY VISIONARY

Davidson has been the principal architect of the company's growth since 1957. Guardian has grown from a local business, consisting of a single glass fabricator to an international operation with manufacturing and fabricating facilities on four continents serving thousands of customers in dozens of countries. Guardian now has over $1.4 billion in annual revenue, and employs over 10,000 people worldwide.

Its product line now includes raw float glass; patterned glass; heat strengthened and fully tempered safety glass; laminated windshields and tempered glass for vehicles; insulating and reflective glass for residential and commercial construction; and glass mirrors.

Guardian subsidiaries include Guardian Fiberglass, Inc., a manufacturer of fiberglass insulation, and Windsor Plastics, Inc. a manufacturer of injection molded plastic products. A Guardian affiliate, Optical Imaging Systems, Inc., manufactures active-matrix flat panel displays for computers.

Guardian has been based in Michigan for over sixty years. In early 1995, its world headquarters moved from Novi to a new building in Auburn Hills.

PERFORMANCE

THE SIMPLE FORMULA FOR SUCCESS THAT HAS DRIVEN MICHIGAN'S LEADING ELECTRICAL CONSTRUCTION FIRM SINCE 1952.

Foresight.

Working to be the best requires commitment, drive and a vision.

Being the best means staying ahead of the game -- having the foresight to anticipate current customer needs and future needs. For well over 50 years, the Guardian Automotive Group has delivered innovative, first-quality glass systems to the automotive industry.

Guardian has 38 manufacturing operations throughout the world plus research and development facilities in the U.S. and Luxembourg. These resources assure our customers of consistently high quality products delivered on a timely basis.

The secret to our success is simple -- teamwork. We work closely with our customers -- through every phase of the design and manufacturing process. We listen to their needs and requirements and strive to deliver a product on-time that not only meets design specifications, but exceeds them.

If these are the kind of attributes you look for on your team -- look to Guardian for Global Automotive Vision.

Global Automotive Vision

GUARDIAN
A Company of Vision

Guardian Automotive Products Group

Guardian Automotive Sales Offices:

Auburn Hills, Michigan
(810) 340-1800 Fax: (810) 340-2375

Upper Sandusky, Ohio
(419) 294-4958 Fax: (419) 294-3264

FORD

HONDA

HONDA-UNITED KINGDOM

TOYOTA

Worldwide Facilities
Integration for the Automotive Industry

- 2,746 Worldwide Technical Staff
- Overseas Greenfield Experience
- 27 Years Presence in the Far East
- Serving the Industry since 1914

Turner
Automotive
Group

The Lathrop Company

Turner Construction Company

Universal Construction Company

Turner STEINER

Auburndale

HONDA

FORD

CHRYSLER

GM

That's The Break(er)s

Square D (for Detroit) switched name over customer recognition

by Tom Yates

It's Monday morning and you and your spouse are getting ready to do battle with the coming week. In the kitchen the coffee maker, microwave, waffle iron and clock radio are busily working away. A hair dryer is plugged in the overworked circuit and the world grinds to a halt; a circuit breaker has tripped. Today that means a visit to the breaker panel where you scan the ranks of breakers for the one out of line. Finding the miscreant, you reset it — and move the hair dryer back to the bathroom.

Ninety years ago, if your home were one of the newly electrified buildings, you probably would have had a made-of-iron electric iron, one or two outlets and the kitchen lights on the same circuit. When a fuse blew then it meant a trip to the fuse box with insulated gloves, pliers and/or a screwdriver and a fuse link. You'd have to pull the main fuse to kill the power to the fuse, remove the ends of the old fuse, install the

TECHNOLOGY FOR PEACE OF MIND...

In the dark of night, it's comforting to know that you — and only you — can open your car door and slide behind the wheel.

That's why at United Technologies Automotive, we have developed and designed a patented encryption authentication technology into our remote entry systems. Would-be thieves can't record the codes — or predict them. Which makes it tougher to break into your car. And that means peace of mind for you, and your customers.

At UTA, we're working hard to apply advanced technologies to meet the needs of you and you[r] customers. That's why almost every car and light tru[ck] in North America and Europe, and a growing number [in] Asia, feature our components.

UNITED
TECHNOLOGIES
AUTOMOTIVE

new fuse link and put the main fuse back in — considerably more complex than just flicking a circuit breaker switch.

The Square D company, which began life in Detroit in 1902 as the McBride Manufacturing Co., made life easier for homeowners in 1936 when it introduced the first residential circuit breaker panel. Until that time breaker panels were reserved for commercial or industrial use. Homes used either cardboard-and-brass cartridge fuses or glass screw-ins. In either case the homeowner had to remove the bad fuse and replace it with a new one — and that's assuming someone remembered to buy new fuses after they used the last one. If not it meant borrowing a fuse from a less vital circuit. If the electrical problem hadn't been corrected the home owner was rewarded with a pop and puff of smoke as the fuse blew again. The Square D circuit breaker solved the no-fuse problem. And virtually every new home built today uses circuit breaker panels instead of fuse boxes.

For engineers, Square D equipment is even more familiar. Every piece of small industrial or commercial equipment and small electric motors use a knife switch to control power flow. In the early part of the century, equipment knife switches were mounted on the wall near the equipment. The blade and contacts were exposed and people often come in contact with them, with results that ranged from irritating to fatal.

SQUARE D-NOTATION DERIVATION

Square D's answer to that problem not only saved lives, it gave the company its name. By 1905 the McBride Manufacturing Company had become the Detroit Fuse and Manufacturing Company. It was Detroit Fuse that introduced the "Iron Clad" safety switch, enclosing the blades and contacts of the knife switch in an iron box.

In 1915 the company introduced a new stamped-steel version of the safety switch. On the cover of the new switch was the letter "D" inside a decorative square, the new trademark for Detroit Fuse and Manufacturing Company. In a short time contractors were asking for the "Square D" safety switch. The name became so popular — and synonymous with the company — that in December 1917, when the fuse business was split off from the manufacturing side and sold, the company name was changed to Square D.

After the name change, the company began to expand, with sales by 1919 exceeding $1 million for the first time. Square D became an important player in the electrical equipment business as manufacturing for appliances and autos boomed. In the 1920s , even Thomas Edison specified Square D safety switches for his Florida research lab. The 1920s also saw the company expand into international sales.

Since then Square D has been a consistent innovator in the electrical equipment field. Its 1925 Saflex power panel made control of industrial equipment simpler. Its plug-in circuit breakers, introduced in 1951, made installation and replacement of breakers easier. The company also introduced the industry's first two-pole ground fault interrupter in 1973.

Stepping into the computer age Square D introduced the world's first microprocessor-based welder control system and the SS/Max-20 programmable controller in 1978. Today the company is a worldwide supplier of equipment to control electricity and electrical equipment. From push buttons to infrared thermometers to busways to factory automation systems, Square D equipment is known and used around the world.

Between 1920 and 1980 the company bought or acquired 24 other electric equipment companies, sold several and merged with other electrical equipment companies three times. Beginning in 1955, Square D began decentralizing its facilities, then still concentrated in Detroit. By the summer of 1959 it ceased manufacturing operations in Detroit entirely and a year later moved its headquarters to Park Ridge, Ill. In 1991 Square D was acquired by European investors, Groupe Schneider, and company headquarters is now in Palatine, Ill , a Chicago suburb.

Today other manufacturers make circuit breaker panels and safety switches. Still, in many trades, the term "Square D panel" or "Square D box" is generic for any circuit breaker panel or safety switch, the terms being passed down from journeyman to apprentice.

Although the Square D Company no longer is based in the city which gave it its unusual name, its heritage hints that — but for chance — Detroit might be known today as "Circuit Breaker City."

Fitting Out The Factories

*New manufacturing equipment
turns sound of plant floor
to a hum from a roar*

by Gary S. Vasilash

**Top: *This scene in a Fisher Body plant in 1918 is
typical of the labor-intensive production methods
in the first decades of the industry. Workers fas-
tened sheet metal skins over wooden frames, largely by hand. (General Motors)***

Try to imagine a factory without
machines powered by their own electric
motors — the manufacturing scene
before Thomas Edison's generating
plants powered up industry after the
mid-1890s. Like trying to imagine
Detroit without the automobile, it's nearly impossible.

Take a walk into any contemporary automotive
production facility, be it a car assembly plant, where
robots wielding spark-emitting spot-welding guns
transform sheet metal skeletons into durable, reliable
body structures, or even an engine plant, where blocks
of cast iron or aluminum are whittled away into the
bodies of high-performance engines by cutting tools
that may even be coated with diamond. These facilities
are clean, well-lighted places. Even if you go to the
stamping areas, where hundreds of tons of force are
applied to shape materials, you may be surprised that
it isn't as noisy as you'd think it would have to be, considering the metal-against-metal impacts.

Although some might imagine that high-tech facilities are more indicative of computers or aerospace
industries, any modern car plant has enough electronics, controls, computers, and communications equipment to hold its own with other more seemingly
"space-age" operations.

If we were to be taken back in time 100 years ago,
we would find quite different circumstances. We would
find workshops that might bring to mind something
out of Charles Dickens. There were no electric motors
powering individual machines. There were, instead,

large power-generating devices — coal-fired steam engines usually — that were used to rotate a shaft that would stretch across the working area. From this shaft would descend leather belts to drive machine spindles.

Architecturally, many of these workshops had saw-tooth-shaped roofs with windows (at best) or other means to let natural light in. Otherwise, the feeble attempts at interior illumination were barely adequate.

BANG-BANG! YOU'RE DEAF

When you take one piece of metal and bring it into contact with another at high force and high pressure, with one of the pieces in motion and the other stationary (think of a turning machine, or lathe, where the part rotates and the cutting tool is stationary; think of a stamping press, where the ram descends on the flat piece of metal that is located above a stationary die), the results include noise that can startle even the hard of hearing. Adding that tremendous din to the picture, the factory environment of yore wasn't the kind of place that any of us would like to spend much time in.

Of course, if we go right to the roots of automotive production, up at least until 1899 and in a plant built in Detroit by Ransom E. Olds (where a static assembly line was pioneered), we would discover that many of the vehicles being produced were done so by people who were making products that tended to be more craft- than mass-production — things like bicycles (Duryea) and carriages (Durant), so the size of these facilities wasn't what is ordinarily found today.

There is a tendency, when considering the technical accomplishments of the past, to think not how far we've come but how — pitifully — little our ancestors knew by comparison. When considering the development and implementation of machinery, equipment and systems in automotive factories during the past 100 years, one must be more surprised at how comparatively advanced were the pioneers who built the infrastructure of this industry , if not in detail, then certainly in imagination, aggressiveness and approach.

This is, perhaps, most evident in the

developments engendered by Walter E. Flanders. He promoted efficient manufacturing operations that were departed from as the century progressed and are now being put back in order.

For example, it is now well known that efficient parts production, whether in machining or assembly, is performed by having the equipment arranged in a sequence. This is often called "one-piece flow," and there are some knowledgeable people who might explain that this development is one that has been adapted from the practices employed in Japanese manufacturing companies. Yes, that may be the case. Many U.S. manufacturing facilities were — and to a fair extent still are — arranged in departments so that there are milling machines in one area, turning machines in another, grinders still somewhere else, and so on.

Yet Walter Flanders, in the first decade of this century, helped Ford Motor Company establish the manufacturing system for the Model N engine at the Bellevue plant, wherein machines were placed in sequence, not segregated by type.

What's more, today there is a push to have suppliers work more closely with the auto manufacturers. One form that this closer relationship takes is in the issue of handling and delivering parts. Although the rules have certainly changed since 1906, when Flanders drafted a policy statement for Ford, he was certainly on the right track when he indicated that it was important (a) to have long-term agreements with suppliers and (b) have those suppliers be responsible for carrying the inventory for the automaker. (A significant difference is that whereas the idea today is to have the

suppliers build the parts and components based on a real-time need, Flanders thought that the carmaker ought to have a 10-day supply of parts on hand — of course, transportation wasn't what it is today, either.)

The importance of suppliers in the auto industry seems to be increasing today, as automakers are working toward having the suppliers do more for them. But this, too, is almost a case of "everything that's old is new again." Back in 1913, when the Model T was being cranked out, parts and materials from outside suppliers accounted for 84 percent of the vehicle's cost.

First 'Flow' Feat

Although a general notion exists that the moving assembly line is a development of Henry Ford (and if not by him, then certainly by one of his key associates, such as Charles Sorensen or Oscar C. Bornholdt, both of whom were instrumental in the development of Ford's manufacturing capabilities in the developing stages), it seems that there was a firm called the EMF Company that pioneered flow production in 1908. This company was managed by two men who had worked for Ford, Max Wollering and — here he comes again — Walter E. Flanders. Between 1908 and 1911, this company produced some 23,000 vehicles. The company was purchased by Studebaker in 1911, the year that a mechanized chassis line was installed: it employed two motor-driven elevator drums and cables; hooks pulled the front axle through the 30 or so stations. It helped increase output to 100 chassis per day.

The Ford Highland Park plant's chassis line was installed in 1913. That year was something of a breakthrough year for the implementation of mechanical lines in automotive factories: in addition to the chassis operation, the lines were also implemented in engine assembly and the foundry. One of the more notable assembly lines started up in May 1913 at a Ford plant for flywheel magneto assembly. Prior to the line, one worker could produce on the order of 35 to 40 magnetos in a nine-hour day; once it was up and running, a worker (there were 14 of them on the line) manufactured an average of 55 during an eight-hour day.

What is also interesting about Walter E. Flanders is what he was doing for a living when he came to Ford Motor Company: he was a machine tool salesman representing three firms, Potter & Johnson, Landis Tool Company, and Manning, Maxwell & Moore.

"Undoubtedly, automotive production technology established the pattern for technical change in the modern mass production industries through the twentieth century," writes Stephen Meyer, III in The Five Dollar Day: Labor and Management and Social Control

in the Ford Motor Company. One of the issues that had a profound effect on what equipment was implemented in facilities was the skill pool of those who were available to run the factories that were developed in the early years of the 20th century. The auto industry was not the first mass-production industry. In 1913, Oscar C. Bornholdt, a machine tool expert at Ford, compared the arrangement of sequential machinery in an auto plant to that in food canning operations; noted historian of technology Siegfried Giedion, in his Mechanization Takes Command, argues that the assembly line's direct forerunner was the hog slaughterhouses of Cincinnati established in the 1860s.

There were other types of goods — from rifles to bicycles — being produced in a mass model by the time the auto industry was still in its early developmental stages. Stephen Meyer argues that even though there was mass manufacturing occurring in these other industries, there was still a significant amount of hand work being performed by skilled mechanics in order to make the parts fit; it was not just a case of part interchangeability (Eli Whitney notwithstanding).

Taylor-Made

Skilled craftspeople were not available in the numbers needed by the burgeoning auto industry. One consequence was, the "principles" of what was called "scientific management" as devised by Frederick W. Taylor were instituted in auto factories (i.e., managers figured out what needed to be done in terms of tasks; workers were then required to do these simple, specific, discrete tasks — something that auto manufacturers of today are doing their best to move away from).

The other was that machinery and equipment were specifically created (a) to permit less-skilled people to do tasks (for example: jigs and fixtures that permitted workers to load and unload parts and the machines to do the manipulative work (echoes of this early work can be heard today when people talk about poka yoke, or mistake-proofing, for part loading) and (b) specifically to do jobs at relatively high cycle rates.

The relationships between the auto manufacturers and machine tool makers can be traced back as early as the development of the Ford Highland Park plant, which opened in 1910. Ford's Charles Sorensen figured that it would be efficient if people within Ford who were most familiar with the work designed the necessary machine tools to accomplish it, then built prototypes. The actual manufacture of the machines would be performed by the companies who specialized in machine building. This relationship exists to this day. One of the earliest suppliers of machine tools to the auto industry, for example, is the Ingersoll Milling Machine Company of Rockford, Illinois.

Ingersoll sold its first machine — a horizontal-spindle mill — to Ford in 1908. This machine had application in other industries besides automotive, of course. But by 1914, as the needs of the auto industry became greater, Ingersoll developed a machine that could continuously machine engine block castings. The operator would load and unload fixtures. The machine would do the rest. As the years went by, Ingersoll developed what are known as "special" machines for Ford (and other automakers). The machine tool maker even helped Ford tool up for aircraft production during the 1940s. In 1985 Ford and Ingersoll began collaborating on a new concept for automotive parts machining, a piece of machinery now known as the High Velocity Machining Center. This machine tool has the capability of moving its spindle at 3,000 inches per second and promises to replace multiple machines — perhaps even transfer lines — because of the speed at which it can perform its work.

In a very real sense, the early Ingersoll machine and the high-velocity machine can be considered matching bookends in that they are essentially stand-alone machines (although they did/do offer the possibility of being integrated with other equipment to form "systems"). In between the two are an array of "transfer lines." What the assembly line was to putting parts together, the transfer line was to machining parts. Essentially, a transfer line is a device where parts, such as an engine block or head, are fixtured in place, and then are transferred, in lock-step, through a series of "stations," at which machining operations (drilling, boring, milling, facing, etc.) are performed. The speed at which the transfer line is moved is fundamentally based on the longest machining cycle in the entire sequence. But the transfer line — built by a variety of companies, including Ingersoll, and The Cross Company, F. Joseph Lamb, and many others — ushered in the wherewithal to perform mass production, as the work is being performed automatically, without human intervention. Even today, transfer lines are in operation and being built for the production of parts where little change is anticipated and the volumes are high.

Generally, as the century is reviewed, it is easy to conclude that there were but a handful of profound innovations in the development of machine tools and equipment. It was mainly an issue of refinements. Admittedly, it may be that some developments — such as the contour bandsaw by DoALL in 1935 or the Bridgeport milling machine in 1938 — are just parts of the industrial landscape that are now taken for granted,

despite their great impact on auto production.

THE OLD GRIND

Consider grinding. It's been with mankind, in one form or another, ever since a harder stone was rubbed against a softer one. The past 100 years are replete with developments, enhancements, to the technology.

The Norton Grinding Company provided the auto industry with a crankshaft grinder back in 1903, the same year that A.B. Landis introduced a device that permitted connecting rod pins to be automatically loaded for grinding. In 1933 abrasive belt grinding was introduced, a process that was enhanced by the discovery of the electrocoating process, which made the abrasive grains stand on end during the manufacture of the belts. Ex-Cell-O developed a precision thread grinder after five years of work that, after it was introduced in 1935, made possible the mass production of precision threads. In 1947 Micromatic Hone Corp. brought out the Micromatic Hydrosize, a honing system for cylinder bore sizing and finishing. In 1956 National Broach & Machine brought out a process called "gear honing" to improve the teeth on spur and helical gears. In 1957 General Electric Co. changed many of the parameters in grinding with the commercial introduction of its synthetic diamonds; GE's Research Laboratory announced that same year that it had developed a synthetic material called Borazon, cubic boron nitride, which was to find application in grinding wheels and cutting tools. In 1972 there was a flurry of developments from companies including Babcock & Wilcox and Brown & Sharpe: these companies came out with trichoid grinding machines in anticipation of the GM Wankel program, a road not taken.

And on it went, through the years. A new machine. A new material. Today the greatest advances in grinding machines tend to relate to microprocessors for controlling all of the parameters involved in producing parts. But, fundamentally, it is rubbing a harder material against a softer one.

There are a handful of 20th century developments that have had massive effects on the way automobiles (and other products) are manufactured. Although the following list is short, and leaves out important breakthroughs in materials and metrology, pressworking and painting, these have certainly been agents of change:

• Numerical control. John T. Parsons, in Traverse City, Michigan, was building munitions for the military during the 1940s. He was also producing helicopter blades. This requires complex contour machining. He started thinking about how to apply math-based data to machine contours more effectively. By 1948, he was

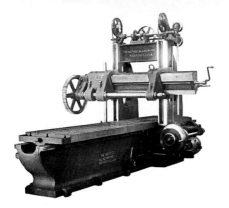

putting the information on punch cards. He called it "cardamatic milling." In 1949, Parsons received a $200,000 contract from the U.S. Air Force Materiel Command to build a machine that could be directed by punch cards to machine integrally stiffened sections for aircraft. Parsons, in search of assistance, went to the Servomechanisms Laboratory of the Massachusetts Institute of Technology. In September 1952, a formal demonstration of numerical control, as it became known (according to the man who ran the project at MIT, they simply picked a name that stuck), was presented. Since that time, the control of machine tools — for all applications — has never been the same. It's gone from punched cards, to punched tape, to magnetic tape, to direct input, to floppy disks in a short number of years. Regardless of the input system, the handwheels once carefully turned by machinists have given way to electronic control.

• The machining center. Although many people think of automotive machining in terms of transfer lines and the great companies that built them — Cross, Ingersoll, Lamb, and Snyder among them — a development of 1958, the machining center, is likely to change the model of machining in auto manufacturing. It was introduced by Kearney & Trecker, and was designed to mill, drill, tap, and bore. This multifunctional machine was fitted with a tool-changing device operated under numerical control, so manual intervention wasn't required during a cycle to move from one operation to another. The consequence of this is "flexibility" — perhaps even "agility" — in production operations. So while the traditional transfer line consists of a sequence of fixed stations that do one thing (mill or drill or tap or bore), the machining center can do all of them in a single setup. As shorter runs and modifications to existing products become the norm, the machining center will become more prevalent.

• The industrial robot. The industrial robot's birth — to use the sort of anthropomorphic term characterizing the equipment throughout its existence — goes back to 1961. The generally accepted father is Joseph Engelberger, whose "Unimate" robot promised to release humans from dull, dangerous and dirty work. The first installation of a robot was in a GM die casting operation. Since the early days, the auto industry has been by far the leading customer for these industrial

tools. The primary application has been in bodyshops, where electrically driven machines (the original Unimates were hydraulic) are capable of swinging the heavy spot welding guns with seemingly no effort. One interesting aspect of robots during the past 30-plus years is the way they have evolved. An original contention was that a robot is a "multifunctional device," something that was capable of doing all sorts of tasks. By investing in a robot, it was argued, a company was actually investing in a piece of machinery that could weld, paint, machine load, assemble. Its evolution has been one characterized by specialization, so there are now robots that paint and robots that weld, and although it is possible that one could be equipped to do the other's task, the likelihood is about nil.

• The programmable controller. This digital brainchild of Richard E. Morley was first applied in the GM Hydra-matic plant in Ypsilanti in 1969. At issue was a better way to control, in relative real time (as distinct from the way computers buffer information before processing), the massive pieces of machinery on the factory floor so there would be better, more precise synchronization than could be attained (at least not without excessive difficulty and expenditure) with relays. The Modicon device has launched an entire array of industrial controls that have not only changed the model of the ways cars are built, but have even made their way into applications such as controlling the traffic lights that pace the cars on the road.

• The laser. This is one of those technologies that has been long in coming, with its full promise in auto production likely to be ahead of us. In 1960, a scientist at Hughes Aircraft announced a source of coherent light. By 1962, companies including Westinghouse, Raytheon and General Electric were working hard to develop the laser for industrial applications, such as drilling and welding. Early lasers used crystals such as ruby. Then they were built with a gas medium, such as CO_2. Meanwhile, better crystals were developed, particularly those based on neodymium-doped yttrium aluminum garnet (Nd:YAG). Today there is coexistence between the two types. CO_2 lasers, which are tremendously powerful, can be found in great number at transmission plants, where they are used for both hardening and welding. Nd:YAG lasers have the advantage of operating at a wavelength that can pass through optical fibers. This means that working in combination with a robot, the Nd:YAG laser can be set up to perform welding or cutting operations without the difficulties involved in aligning and maintaining mirrors, as is necessary for CO_2 lasers. Look for more Nd:YAG installations in body shops in the years ahead.

A continuous caster at National Steel's Great Lakes Division in Ecorse pours molten steel in 1988. (Detroit Free Press)

Ready Ore Not

Raw material processing: setting the stage for heavy industry

by Bob Casey

Detroit's roots in materials processing lie deep in the 19th century. In 1856 in Hamtramck Township, on the Detroit River near the present site of the Belle Isle Bridge, Dr. George B. Russel built the first American iron blast furnace west of Pittsburgh. It remained in operation until 1905.

The first American steel produced by the Bessemer process was made at Capt. Eber Brock Ward's Eureka Iron Company in Wyandotte in 1864. Unfortunately Ward was unable to capitalize on his development, and Pennsylvania became the center of Bessemer steel production. Three major steel companies continue the legacy begun by Russel and Ward: Great Lakes, McLouth, and Rouge.

Great Lakes Division of National Steel traces its origins to the Detroit Iron & Steel Company, which erected blast furnaces on Zug Island in 1902. Despite its name, Detroit Iron & Steel produced only iron. By 1929 the works were owned by M. A. Hanna Company, which also owned ore mines, coal mines, and lake freighters. In 1929 the Hanna properties were combined with Wierton Steel Company of Wierton, W. Va., and Great Lakes Steel, which was rolling sheets from steel purchased from other sources. The new company, called National Steel, set about constructing an integrated steel mill on 275 acres in Ecorse. Six open hearth furnaces, plus blooming, hot strip, and bar mills, turned pig iron from Zug Island into semi-finished steel. Over the years the complex has been enlarged and modernized, with the addition of Bessemer converters in 1946, an 80-inch hot strip mill in 1961, basic oxygen furnaces in 1962 and 1970, and a continuous slab caster in 1977.

The 1970s and 1980s were hard times for steel mills, and Great Lakes did not emerge unscathed. By 1981 its capacity was down to 4 million tons per year, from a peak of 6 million tons. Recent years have brought outside investment from the Japanese steel maker NKK Corporation, which now owns 75 percent of National Steel.

SCRAP TO SCRIP

McLouth Steel Products (formerly McLouth Steel Corporation), was the brainchild of scrap dealer Donald B. McLouth. McLouth started small in 1934, rolling steel strip from slabs supplied by other companies. The purchase of government surplus electric furnaces after World War II allowed McLouth to produce its own steel from scrap. Rising demand and shrinking scrap supplies pushed the company toward integrated steel production in the early 1950s. Taking a chance on new technology, McLouth became the first American company, and the fourth in the world, to adopt the basic-oxygen furnace developed in Austria. In 1964 the company again chose a relatively untried technology when it became the first American integrated producer to base all of its production on continuous casting.

Adoption of new technology was not enough to insulate McLouth from the changing climate for steel makers. By 1981 the company had filed for bankruptcy under Chapter 11 of the Federal Bankruptcy Act. A group headed by Cyrus Tang bought McLouth in 1982, and in 1988 employees purchased 85 percent of the company under an employee stock ownership plan.

Henry Ford's Rouge plant was designed to take in raw materials and send out finished cars. Key to this concept was the production of iron and steel. In 1921 Ford built the first of three blast furnaces, along with coke ovens and other auxiliary equipment. Steel was first made at the Rouge in 1923 in electric furnaces. The first open hearth furnaces went on line in 1926, followed by a hot strip mill in 1935. The decades after World War II saw many new additions: basic oxygen furnaces in 1964; a new hot strip mill in 1974, new electric furnace in 1976; a continuous caster in 1986. The changing economics of steel production in the 1980s brought changes that would have amazed Henry Ford. In 1982 Rouge Steel became a wholly owned subsidiary of Ford Motor Company. Four years later Rouge entered into a joint venture with USX Corpora-

tion. Together they operate Double Eagle Steel Coating Company, the world's largest electrogalvanizing facility. Finally, in 1989 Ford sold Rouge Steel to Marico Acquisition Corporation. Today, surrounded by the Ford Motor plant, Rouge Steel is an independent company.

Processing of copper and brass was another long-time Detroit industry. By the mid 1850s Detroit and Lake Superior Copper Company was the largest copper smelter in the U.S. By the end of the 19th century, however, actual smelting had moved to the Lake Superior mines themselves, and Detroit establishments confined themselves to secondary processing.

TOP BRASS

Detroit Copper and Brass Rolling Mills was founded in 1879 to roll copper ingots into sheets, bars, and rods. The company expanded into the production of brass ingots, first in crucible furnaces, and by the 1920s, in electric furnaces. In 1927 American Brass Company, a subsidiary of Anaconda Copper Mining Corporation, bought the Detroit works, and operated them until they closed in 1957.

The Michigan Copper & Brass Company began as a small roller of copper and brass in 1907. By 1928 the factory next to Fort Wayne on the Detroit River had expanded into brass and aluminum extrusions and bronze and aluminum castings. That year it was one of six companies that merged to form Republic Brass Corporation. In 1929 the new firm became Revere Copper & Brass, Inc.; the Detroit operation closed in 1981.

Michigan Smelting & Refining Company was established in 1895 as a scrap metal dealer. It expanded into the production of copper and brass, and built a new factory in Hamtramck in 1912. Here the company, in co-operation with the Bureau of Mines, pioneered the development of electric furnaces for non-ferrous melting. Products included brass and bronze ingots, brass billets and slabs, solder, babbitt, lead and tin pipe, brazing spelter, and die castings. The company closed after World War II.

Today the tradition of non-ferrous processing is carried on by smaller firms. Arco Alloys, established in 1938 by Harry Arnow and Leonard Cohen, produces zinc alloys. Originally relying on scrap zinc from automobiles, the company now uses virgin zinc, aluminum, and other metals to produce alloys for die castings. Equitable Metals Corporation began in 1953 as Sollish Scrap Metals, a dealer in non-ferrous scrap. Owner Henry Sollish began refining precious metal scrap in 1964, and in 1967 the business was incorporated as Equitable Metals. It refines gold, silver, palladium, and platinum.

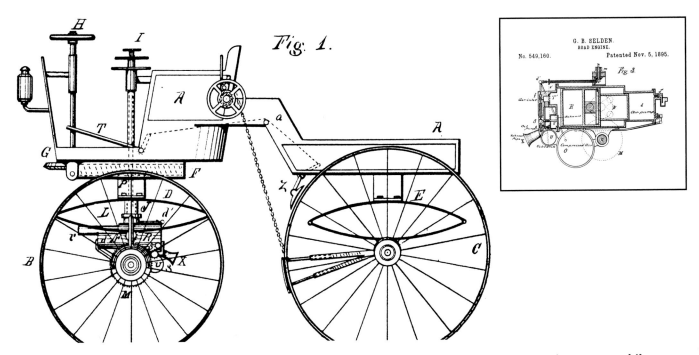

Attorney George B. Selden's 1895 patent for a horseless carriage, and its engine (inset). (American Automobile Manufacturers Association)

Patents Pending

Industry's needs were the mother to many inventions

by Tom Yates

In 1895 Detroiters, like people today, had to wonder what was left to invent. They already had the electric light, electric motors and the telegraph; even the telephone was 20 years old. Thomas Edison received a patent for his phonograph in 1895. Still, the Patent Office kept issuing patents — 22,057 that year — so the end of progress wasn't in sight.

One patent that year had a great effect on Detroit, even though it was issued to an inventor in Rochester, New York. According to the U.S. Patent Office, Letters Patent No. 549,160 for a "Road Engine" was issued to George B. Selden of Rochester, New York. To realize how great Selden's vision was, you have to understand he filed his first "Road Engine" patent application on May 8, 1879. According to Dan Kirschner, of the American Automobile Manufacturer's Association, Selden held off the patent process as long as possible so it would protect his concept longer. He foresaw that the "Road Engine" would become an important part of American society.

Despite the fact that the automobile industry was little more than a gleam in the eyes of some futuristic zealots in 1895, people in Detroit and Michigan were inventing things that would soon make it a reality.

A quick scan through the Patent Office Gazette volumes for 1895 shows that trolley car and rail-related inventions were popular for Detroit area inventors. Trucks (the sets of wheels under trolleys and rail cars) were being invented and improved. Bumpers to stop the cars from rolling were popular too. Since great minds seem to think in the same relative paths it seems only logical that another Detroiter should be working on the other half of the streetcar business, a street car switch adjuster.

At the end of the l9th century industry was building up and power was needed to keep factories running. Steam was a popular source of power and steam boilers, engines, pumps and other equipment were being developed then. Again people in the Detroit and Michigan area were busy in their shops and laboratories. A number of boilers were invented by Detroit area dreamers.

July 23d 08
Ignition System

C.F. Kettering

Witnesses

This crude diagram was the beginning of a revolution in the auto industry. It is Charles F. Kettering's original drawing for the patent for the electric automobile self-starter and ignition system later introduced on the 1912 Cadillac. Before that, cars had to be hand-cranked to start. (General Motors)

INSPIRED BY INDUSTRY

Southeastern Michigan inventors seemed preoccupied with industry. Along with car trucks, street car track switch adjusters and boilers they were inventing machines and the equipment to make them more efficient or maintain them. Patents were issued to Michigan inventors for a quill cutter (the metal tubing type, not the feather), a purifier, separator and grader, a tap and bushing and the mundane, but necessary, subject of waste disposal with a waste valve.

As the automobile became more popular and important to the southeastern Michigan area, automotive-related inventions became prevalent. The year 1929 saw an automatic transmission patented by Detroiter Charles Short with rights assigned to GM. The unit only shifted up automatically, down shifts were manual. But that transmission was the precursor to GM's Hydra-Matic transmission introduced by Oldsmobile in 1939.

While GM receives credit for introducing the first all-steel production automobile body with its 1935 "Turret Top" models, others were working on the concept too. Shortly after GM began working on the automatic transmission, the Budd Company received patent rights to all-steel automobile bodies in 1932 and '33. While the Budd Company was based in Philadelphia, the bodies were used by many Detroit manufacturers. In 1933 Amos Northrup, of Pleasant Ridge, was assigned rights to an all steel body design. Northrup assigned rights to the Murray body company. One year later Harley Earl received a design patent for what became the basis for several late-30s GM products.

During the WWII years and shortly after, fewer automotive patents were issued. But in 1956 Harry Keeler of Detroit received a patent for an "Automatically Controlled Braking Apparatus" (ABS system). Keeler's braking system reduced the tires' tendency to skid on snow or slick pavement. Keeler wasn't alone in his thoughts of braking improvements. Paul Kelm and Robert Treseder assigned their patent for an "Anti-skid Hydraulic Brake System" to GM. Braking and skid control weren't just concerns in the latter part of the 20th century. Back in 1911 Robert Moore, a Chicago inventor, received a patent for an "Anti-Skidding Device for Automobiles." While short on both the electronics and hydraulics of today's ABS units Moore's at least made an effort to control skids.

Where ABS was intended to prevent crashes, inventors in the '60s and '70s began to address the inevitable results of accidents One of the first air bag patents, for a design by Michigan inventors Sidney Oldberg from Birmingham and William Carey of Farmington was assigned to Eaton Yale & Towne in 1968.

ANTIPOLLUTION REFINEMENTS

The passing of antipollution statutes in the mid-1960s brought about changes in the ways engines were controlled. In 1978 a "Spark Timing Control System for an Internal Combustion Engine" (or electronic ignition system) was patented by Robert Masta from Ann Arbor.

Masta's patent rights were assigned to GM. Shortly after, in 1979, Motorola received the patent rights to Todd Gartner and Robert Valek's invention of an "Electronic Ignition Timing System Using Digital Rate Multiplication."

Detroit and Michigan are still active in the inventing field. Between 1969 and 1994 more than 65,000 patents were issued to area residents. In 1994 113,268 U.S. Patents were issued, with Michigan inventors accounting for 3,295 of them. In 1895 the Patent Office was issuing patent numbers in the mid 500,000 range. Today they're well into the five million zone.

In the latter half of the 19th century fiscal conservatives called for the closing of the U.S. Patent Office. It was outdated "because everything had been invented." One hundred years later, when tempted to assume the same, we are often reminded that we're still just spooling up for the future.

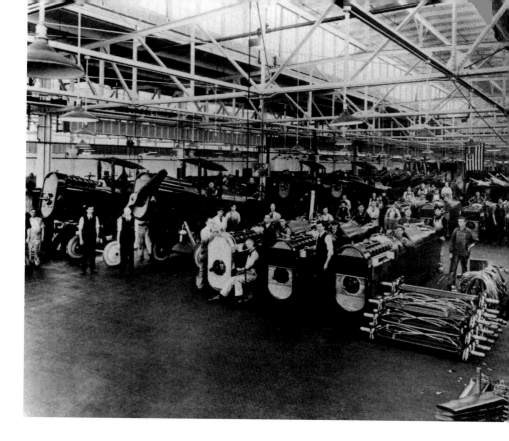

"Liberty" planes under construction on government contract at the Fisher aircraft facility on West End Avenue, Detroit, 1918. (General Motors)

From Wheels to Wings

Move to aircraft development, manufacture is natural evolution

by Paul A. Eisenstein

DEARBORN — The summer air hangs heavy and still, punctuated by the distant backfire of a Fordson tractor and the bellow of a cow. If you listen carefully, though, you'll hear faint buzzing, like a swarm of grasshoppers taking flight. As it draws nearer, the gentle drone evolves into a deafening roar, for off in the distance, a bright silver Tri-Motor banks and turns, skimming the farm fields as it glides in for a landing at Ford Air Port.

The old airstrip is gone now, transformed into a high-speed test track, but the Dearborn Inn, the world's first airport hotel, serves as a reminder of the era when Detroit took to the sky. For many of the pioneers who helped put America on wheels, their fascination with machines and motion knew no earthly bounds. And so, in the early years of this tumultuous, inventive century, the Motor City became the center of the emerging aviation industry.

Detroit's interest in aviation took wing soon after the Wright brothers' first flight at Kitty Hawk. After all, the city had the brainpower and mechanical talent — and the money to invest. The list of names linked with the nascent aeronautic industry soon read like an automotive Who's Who: Packard, Hudson, Stinson, Marmon. And Ford. In 1909, 15-year-old Edsel won his father's permission to build an airplane at an old barn on Woodward Ave. just north of downtown Detroit. Powered by a modified Model T engine, the high-wing monoplane made its maiden flight a year later at one of Henry's farms in Dearborn.

It took a war to make people take the airplane seriously. In 1917, the U.S. could field not a single military aircraft, so the mobilization of air power had to start literally from the ground up. Fisher Body, then an independent manufacturer of automotive bodies, tooled up for production. The Fisher brothers began building aircraft in 1917 out of a plant on West End Ave., a factory that eventually became Plant 18 of the Fisher Fleetwood (eventually the Cadillac) complex. But Detroit's biggest contribution to the war effort was, if you will, under the "hood."

"During the war, Detroit became the great center of aircraft engine manufacturing," explains Bob Casey, historian and curator of the automotive collection at the Henry Ford Museum. The workhorse of the Army Air Corps was a modified version of Packard's robust Twin-Six. By the time the Armistice was signed on Nov. 11, 1918, the Buick and Cadillac divisions alone had produced about 2,500 Liberty engines. Thousands more had rolled out of plants operated by Packard, Ford, Lincoln and Marmon.

In peacetime, U.S. aircraft production nosedived — from 14,000 in 1918 to just 300 two years later. Still, the city's business leaders were convinced aviation would rise like the phoenix, and in the coming decade, they formed a succession of ventures, including the Detroit Aircraft Corp. and Aircraft Development Co. Many of these, and their suppliers, were scattered around old Detroit City Airport.

SEMINAL START-UP

Perhaps the most influential of the start-ups was founded by William Bushnell Stout. His failing eyesight kept Stout from graduating college, but it didn't blind his imagination. Stout was building model airplanes even before the Wright brothers' first flight. An avid tinkerer and inventor, Stout had limitless talents for things mechanical, designing anything from toys to motorcycles. In 1916, he hired on as chief engineer for Packard's aircraft division, helping improve the Liberty engine.

In 1919, he established his own company, Stout Engineering Labs. The wood-and-canvas biplanes of the day were slow and cumbersome, and the near-sighted inventor believed a single, metal wing would be far more efficient. Stout demonstrated a metal monoplane torpedo bomber to the military, but a crash landing cost him the contract. His prototype Batwing could have served as the preliminary model for today's Stealth fighter, but the military wasn't ready to bite

back then. A self-styled "imagineer," Stout wasn't much of a businessman. Running short of cash, he mailed a letter off to more than 100 of the nation's leading industrialists. "I should like $1,000," he wrote, "and only promise you one thing: You will never see it again, but you will get $1,000 worth of education." Among those anteing up were Henry and Edsel.

Though Henry Ford wouldn't take his own first flight until 1927 — with Charles Lindbergh behind the stick — the automotive pioneer was ready to bet millions on aviation. He invested in the Stout Metal Airplane Co., and provided land for an experimental airport. Using the Liberty engine, the "Maiden Detroit" made its first flight from Dearborn to Chicago with six people on board. The 133-minute trip launched commercial passenger service. Stout eventually sold the company to Ford, and went on the payroll, as both engineer and manager of Stout Air Services, the Ford airline.

Commercial air service owes a major debt to Ford. One of his most significant and long-lasting contributions was the development of the first effective radio navigation system. By following the directional beacon, a pilot could tell where he was headed. Aimed anywhere within 30 degrees of the tower, he'd hear a steady tone. Drifting off course, he would hear either an "A" or an "N" transmitted in Morse Code. The first beacon went operational at Wright Field in Dayton on Feb. 16, 1927, permitting Ford test pilot Harry Brooks to navigate his way down from Dearborn in a heavy snow. Ford never took a royalty payment for the beacon patent, and a few years later, the federal government took over responsibility for setting up a network of the towers across the country.

FLYING FORD

The need for more power, space and passengers led to the development of the legendary Ford Tri-Motor. With its three Pratt & Whitney radial engines and a wingspan of 70 feet, the "Tin Goose" was able to carry a crew and eight passengers, as well as a load of freight and air mail — ensuring profits for the new airlines — such as Pan American, Northwest and Transcontinental Air Transport, later TWA, popping up across the country.

Ironically, Ford made the same mistake that cost him leadership in automotive manufacturing. As with the Model T, he refused to replace the Tri-Motor with a bigger, faster and more powerful model, and the competition soon passed him by. One of the most successful was the Douglas DC-3, developed by onetime Ford engineer James McDonnell, who left to form his own company in 1932.

There were other failures. In 1917, Ford, Charles F. Kettering, Wilbur Wright and several others proposed a secret weapon they believed would help the Allies win the war. Nicknamed "The Bug," the pilotless drone was designed to carry a 200 pound bomb across enemy lines using a sophisticated autopilot. Then, in 1920, Ford proposed to the government a plan to set up a million-dollar factory to produce zeppelins in Detroit. He even erected a 225-foot dirigible mooring tower on a corner of Ford Air Port. Both ideas were shot down.

Aircraft Development Corp., established in 1920 by Detroit promoter Carl Fritschie, also attempted to secure a federal contract to build metal-clad zeppelins. The company went so far as to build a hangar and airstrip, which still operates, on Grosse Isle. Its prototype ZMC-2 took flight on Aug. 19, 1929, and was used for the next dozen years as a trainer at the military airfield in Lakehurst, N.J., but the explosive crash of Germany's hydrogen-filled Hindenburg scuttled interest in the commercial lighter-than-air craft.

In 1929, the U.S. Navy contracted with Great Lakes Aircraft for 18 primitive TG-1 torpedo bombers. Successor Detroit Aircraft Corporation then developed the TG-2, and by 1938 had built 277 under contract as Navy T4Ms.

Aiming High

Ford's most grandiose scheme did get off the ground — just barely. The Flying Flivver was the pet project of Henry and his chief test pilot, Brooks. It was meant to be an airplane for the masses and Brooks actually used one to commute to Dearborn from his home north of Detroit. The first single-seat, 550-pound Flivver had a 23-foot wingspan, a 16-foot fuselage, with a 3-cylinder, 35-horsepower engine cruising at 85 mph.

Put into operation in July 1926, it's now on display at the Henry Ford Museum. The second Flivver featured a more powerful, 2-cylinder engine designed by Ford, more stable wings and a gas tank large enough to provide a theoretical range of 1,700 miles. But on a distance record run in late February 1928, Brooks was killed when the Flivver crashed into the ocean off Melbourne, Fla. Distraught at the loss of his friend and chief test pilot, Ford grounded the program until a half-hearted effort was made a decade later.

Ford wasn't the only one intrigued by the Flivver concept. It also caught the eye of Alfred P. Sloan, the legendary boss of General Motors. In his celebrated autobiography, "My Years With General Motors," Sloan noted that "it became steadily clearer that aviation was

to become one of the great American growth industries." GM made a series of key acquisitions, starting with the Dayton-Wright Airplane Co., and later adding a 24 percent stake in Bendix Aviation, and 40 percent of the U.S. subsidiary of the legendary Dutch Fokker Aircraft company. (Ironically, Ford had backed away from Fokker a few years earlier, fearing negative reaction to the name associated with one of Germany's most efficiently deadly weapons.)

Meanwhile, Detroit's most successful venture in building aircraft developed independently of the auto industry. This was Stinson Aircraft, a brainchild of the Detroit Board of Commerce which helped incorporate the company in 1926. From the late '20s to 1948, Stinson was America's largest producer of cabin planes.

Over the next few years, the GM aircraft ventures went through a series of changes. Fokker himself was forced out and the unit was merged into General Aviation, which eventually became a part of North American Aviation. In turn, North American eventually sold off several units, including two airlines. Eastern Air Transport eventually became the well-known Eastern Airlines, while Transcontinental changed its name to Trans World Airlines. (Ford Air Services also survives to this day. After it was spun off, it merged into what is today known as United Airlines.) Resettling in Inglewood, CA, North American developed some of the most significant aircraft of the second World War, including the P-51 Mustang fighter and the B-25 bomber, which General Jimmy Doolittle used to raid Tokyo less than six months after Pearl Harbor.

Detroit Delivers

The sneak attack caught America nearly as unprepared as it had been before World War I. And once again, America turned to Detroit, the Arsenal of Democracy. Along with passenger car production, the Flivver was put on hold.

GM's wartime annual reports were thick with the list of weapons it designed or built for other companies, everything from machine guns to incendiary bomb nose cones. There were turret bearings and B-24 landing gear sets. The V-1710, developed by Allison Engineering Co. was the first military engine to top 1,000 horsepower. Later versions generated as much as 2250 hp. Before the end of 1941, Allison's plant in the shadow of the Indianapolis Speedway was already rolling out 1,100 V-1710s a month for use in such planes as the Lockheed P-38 Lightning. By the time production was canceled six years later, more than 70,000 were produced.

During World War II, GM's Eastern Aircraft Division built more of these Grumman Avenger Navy torpedo bombers than Grumman Corp. (General Motors)

A major GM effort was production of Grumman Wildcat F4F fighters and Avenger TBF torpedo bombers, such as the one flown in combat by President George Bush. GM's Eastern Aircraft Division turned out 7,546 Avengers (known as TBMs when built by GM), far more than the parent Grumman's output of 2,291 Avengers. In World War I, GM's military contracts totaled $35 million. In World War II, that shot to $12 billion, with aviation accounting for 42 percent of the total in 1944.

The other major player in the aerial weapons rollout was Ford. It landed a $200 million contract to produce the Consolidated B-24 Liberator at the Willow Run plant near Ypsilanti. The 32-ton precision bomber was capable of flying at 32,000 feet to avoid antiaircraft fire. Charles Sorenson, who had developed the Model T moving assembly line, was brought back into action. Adapting the same mass-production concepts, Ford could roll out one B-24 an hour, and in the process, costs fell from $379,000 to just $216,000 a plane. When B-24 production ceased in 1945, Ford had built 8,685 of the four-engined giants.

The same techniques helped with the Pratt & Whitney 18-cylinder aircraft engine, of which Ford built 57,851 at the massive Rouge complex. At the start of production it took 2,330 man hours per engine. Eventually, that fell to just 905. Ford also produced many of the gliders used in the D-Day invasion at the company's station wagon body plant in Iron Mountain, Mich. The CG-4A carried up to 15 soldiers, while the larger CG-13A could land a troop of 40.

A SHOT IN THE ARM

Detroit's smaller manufacturers also played a role in the war build-up. In 1948, Time magazine noted "The war gave Hudson a second chance. After a 1940

loss of $1.5 million, Hudson netted an average of nearly $2 million a year making antiaircraft guns, 700-cubic inch invasion barge engines and aircraft parts." By the time war production ended for Hudson on Feb. 13, 1946, the company had built more than half the 98-foot fuselage sections, wings and ailerons needed for the vital B-29 Superfortress bomber, as well as wings for the Curtiss-Wright Helldiver and the P-38 Lightning, cabins for the Bell P-63, fuselage sections for the B-26 Marauder and other aircraft components. Its wartime efforts twice earned Hudson the Pentagon's coveted "E" award for excellence.

As wartime production was phased out, the Motor City shifted back to building cars. With the aircraft industry moving out to the West Coast, Detroit began to lose interest. GM sold off its stake in North American by 1948, though Allison continued to produce aircraft engines, notably during the Korean War. But the era of commercial aviation was at an end — with one notable exception.

In 1985, Chrysler Corp. purchased the corporate jet manufacturer, Gulfstream Aerospace, for $637 million. At the time, Chrysler Chairman Lee Iacocca was trying to diversify his company, creating what he described as a "four box" strategy. The acquisition, he promised, would have the added advantage of providing Chrysler's automotive designers access to the latest aerospace technology. But the synergies never surfaced and except for giving Iacocca access to the new G-4 corporate jet, the deal was a dud. In March 1990, Chrysler sold Gulfstream to a consortium including its founder, Allen E. Paulson, for $825 million.

It was the end of an era. Though Bill Stout was still trying to design an ornithopter when he died in 1951, the dream of a Flying Flivver had already been forgotten. Detroit decided its future lay in mass production. Despite the suggestive name of its Hughes Aircraft subsidiary, GM has only the most minimal role in aerospace today, as do the other Detroit manufacturers. While their profits are climbing back into the stratosphere, the Big Three are content to keep their wheels planted firmly on the ground.

The Stinson "105," introduced in 1939, represented the company's entry in the low-cost private aircraft market. It became the basis for the World War II L-5 observation aircraft. (The Detroit News)

Cabin Cruiser

Stinson rose to new heights as Detroit-based airplane maker

by Mike Davis

O vershadowed by the auto companies and forgotten in the later years of the 20th century and the relative decline of private aviation, the tale of Stinson Aircraft Company is one more Detroit story demanding to be told.

Through much of the period between the rise of aircraft manufacture ignited by Detroit-born Charles A. Lindbergh's New York-to-Paris flight in May 1927 and World War II, Stinson was famed as the nation's leading producer of cabin airplanes.

Stinson was founded by Edward A. (Eddie) Stinson, an Alabama native who learned to fly in 1911, participated in a traveling air circus with his brother and two sisters — all pioneer pilots — and served as a U.S. Army Air Corps first lieutenant and flight instructor in World War I.

During the '20s, newspapers were filled with stories of his numerous record-breaking flights and death-defying escapes from crashes. In 1925, encouraged by the Detroit Board of Commerce, which provided him with space in a loft building on West Congress Street, Eddie Stinson began design of a cabin airplane of more modest proportions than Stout's soon-to-become Ford Trimotor.

Blueprints for the plane were executed at the University of Detroit by Peter Altman and Arthur Saxon, who later became chief engineer of the enterprise. The first experimental plane was completed early in 1926.

It was an equal-span, four-place cabin biplane powered by a single 200-hp Wright Whirlwind radial engine.

Following successful test flights of the prototype, and under the guidance of William A. Mara, then aeronautical secretary of the Board of Commerce, the Stinson Aircraft Company was incorporated on May 4, 1926. The new company promptly leased a plant at Northville for production and a nearby flying field for testing and delivery.

The following year, under the direction of chief engineer William C. Naylor, the company brought out its first cabin monoplane for four, six or eight passengers. It was called the Stinson "Detroiter" and became an almost overnight success. The company's business grew so rapidly that in 1928 plans were announced to purchase a plant site near Wayne, to which production was transferred in 1929.

The Wayne plant was located on a new square-mile Detroit-Industrial airfield at Van Born Road and the Pere Marquette railroad line.

Automotive entrepreneur E. L. Cord then came into the picture, acquiring Stinson in 1930 to mate with the Lycoming Manufacturing Company of Williamsport, Pennsylvania, a long-prominent engine maker. Subsequently, Stinson ownership shifted to Convair, and early in 1939 to a company called Aviation Manufacturing Corporation.

In the meantime, Stinson had developed a series of popular cabin planes for both early airline efforts and private, or general aviation, flying.

In 1928, Stinson Model A cabin planes were utilized by Thompson Aeronautical Corporation of Cleveland to inaugurate commercial flights between Detroit, Pontiac and Muskegon, using the new field at Pontiac as a base. Passenger and mail flights to Bay City, Chicago and Cleveland were added by the following year. In 1931, Thompson changed its name to Trans-American Airlines, which in 1933 was merged into today's American Airlines.

The high-performance Stinson Reliant five-place, gull-winged cabin plane was popular among charter services and police departments in the late 1930s. (The Detroit News)

The Stinson Model T was put into service on Northwest Airline's initial flights between Minneapolis and Chicago, and by Century Airlines for routes between Chicago, Detroit and Cleveland. The Stinson T (for tri-motor) was a ten-passenger craft originally designed by the Corman Airplane company, another of the firms merged into Cord, but produced by Stinson.

In 1932, despite the death of founder Eddie Stinson in a crash on Jan. 26 of that year, the Stinson organization replaced the Model T with an improved Model U for airline service, while a smaller Model S was replaced by the Model R which offered a choice of fixed or retractable landing gear. Two years later, Stinson brought out a new low-winged Model A tri-motor airliner with a cruising speed of 155 mph. In 1936, Stinson also introduced a twin-engined model, the six-passenger low-winged Model B (for bi-motor).

BESTED BY WEST

Both the Ford/Stout and Stinson tri-motor airliners were supplanted in the late '30s by the West Coast-based, more advanced twin-engine designs of Douglas, Boeing and Lockheed, all of which had had Detroit connections. Stinson continued as a leader in private aircraft, however.

Perhaps the most memorable of Stinson cabin planes was the radially-engined Reliant, a five-passenger, fixed-landing gear (though handsomely enclosed with streamlined nacelles) monoplane brought out in the mid '30s. Its high wings were distinctively gull-shaped. The Reliant could be equipped with a 400-hp Pratt & Whitney Wasp Jr. giving a speed of 167 mph with a range of 600 miles or a 450-hp Wright Whirlwind capable of 176 mph.

The relatively high-performance and commodious Reliants were popular with police departments. Even the New York City police had both land-based and pontoon-equipped Reliants.

By 1940, Stinson was producing a three-place, Lycoming-powered Model 105 high-wing monoplane which sold for under $3,000 (about the price then of an upper-series Cadillac convertible) in addition to the five-place Reliant gull-wing. Average employment at the Wayne plant was about 400 with 600 at peaks. When a small fire hit the plant in March of that year, 15 planes were in final assembly with 75 fuselages waiting. Sales exceeded a million dollars annually and the firm boasted of being the largest cabin plane manufacturer in the country throughout the '30s.

After securing a $1.8 million War Department liaison plane contract for the U.S. Army, Stinson was bought by Vultee Aircraft, Inc. in September 1940. Newspaper reports speculated that the contract called for delivery of about 100 airplanes, though details were kept secret by the government and Stinson.

The Stinson L-5 observation plane developed for the Army apparently was based on the Model 105. It weighed 2,100 pounds and was powered by a Lycoming 190-hp engine. After taking over, Vultee split Stinson into two components, continuing civilian production at Wayne and building a new plant outside Nashville, Tenn., for the L-5.

In addition to the U.S. contract, the British government contracted for delivery of Reliants, redesignated as the AT-19, for navigation training. As a consequence, employment at Wayne rose to 900 late in 1940. Altogether, during World War II, Stinson produced more than 5,000 L-5 Sentinels and 500 AT-19 Reliants.

Shortly before the U.S. entered World War II, Stinson introduced another three-place civilian model, the Franklin-powered 90-horsepower Voyager, which was similar to the 105 but slightly faster with a top speed of 115 mph. All three — 105, L-5 and Voyager — used opposed-cylinder engines, rather than radials as in the Reliant, and were equipped with wing modifications such as slots and flaps for short landings.

FLYING STATION WAGON

After the war and anticipating a great demand from returning-veteran pilots, Stinson quickly converted to civilian production. The new planes, based on the L-5, were the four-place Voyager and the Flying Station Wagon models, both with Lycoming engines. Despite the fact Stinson had to compete with plentiful, low-cost war-surplus planes, more than 5,000 Voyagers were built. According to a Detroit News article, during 1946 and 1947 half of all four-place planes in the country were assembled by Stinson at the Wayne plant.

But the boomlet collapsed in mid-1947, when production was suspended. Using parts on hand, workers completed 773 planes in 1947 with a value of $3.7 million, indicating a wholesale price of about $4,900 each.

The final chapter came on Dec. 6, 1948, when Stinson was acquired from what was by then Consolidated-Vultee by rival Piper Aircraft of Lockhaven, Pennsylvania, and airplane production ceased at Wayne — and in Michigan — after a run of more than 20 years. Two years later, General Motors' Detroit Diesel Division took over the plant to produce parts for Korean War tanks.

Although Ford Tri-Motors are better known, Century Air Lines flew Stinson three-engined planes, such as this one photographed April 13, 1931 on regular passenger service between Chicago, Detroit and Cleveland. (The Detroit News)

DETROIT STOVE WORKS

Long before it became "Motor City," Detroit could boast of the "largest stove plant in the world," located at the Belle Isle Bridge and East Jefferson. (Donald M. D. Thurber collection)

Stoveworks: A Warmup

Industry laid solid base for auto manufacturing successor

by Donald M. D. Thurber

When the Dwyer brothers, Jeremiah and James, started their small gray-iron foundry at the foot of Mt. Elliott Avenue on the Detroit River in 1864, they displayed sound business judgment and economic sense. They could not know, of course, that theirs was a first step toward the establishing of Detroit as the world's leading center for the forming and shaping of metals, first into stoves, later into automobiles.

The Dwyers enjoyed the economic advantages of their timing and location. Iron ore had been discovered in Michigan's Upper Peninsula in 1844. Immediately, smelting operations began in many locations where there were boat harbors and nearby hardwood forests to feed the charcoal kilns that produced a necessary ingredient in the conversion of raw ore into the pig iron needed by the Dwyers and others.

Limestone went into the smelters too, and that was plentiful and available in many areas of Michigan. Finally, pig iron enjoyed a cheap trip by sail or steam right to the Dwyers' dock. To move its finished products to market, the foundry had both water and rail connections.

Even when the era of charcoal iron was replaced by the use of coke in much larger furnaces located around the southern fringes of the Great Lakes, Detroit's metalworking industries were strategically located, mid-way between the iron ore of the North and the coal and coke of the East and South. The Dwyers' choice of location would be vindicated many times over during the century and more that followed their modest beginning.

The Dwyer foundry turned out its first stove in 1864. It was an attempt to enter the consumer market with a finished product, as opposed to the usual job-lots of castings for other manufacturers and fabricators. The foundry produced one stove a day, which one of the Dwyers would take to local hardware stores until it was sold.

William H. Tefft, a wholesale hardware merchant, was impressed by the ability of the Dwyer brothers and offered the capital to expand their partnership into a corporation, the Detroit Stove Works, and to build the first unit of what eventually became a ten-acre plant on East Jefferson Avenue at Grand Boulevard. It was then in Hamtramck but eventually became part of Detroit.

PRODUCTION HEATS UP

By 1870, six years after the first one-a-day stove, the company produced 16,500 units annually; in 1880, 30,000; and in 1887, 60,000. By then it was the largest stove plant in the world and from 1880 on Detroit was the acknowledged stove capital.

Jeremiah Dwyer was forced by illness to withdraw from the Detroit Stove Works about 1870 but was by no means through as an industrial innovator and pioneer. Recovering his health, he and some other capitalists formed the Michigan Stove Company, which began its rapid growth in 1872. Its plant was at East Jefferson and Adair Streets, a few blocks east of the other stove company. Both factories were served by water and rail.

The Detroit Stove Works made 700 varieties of stoves, under the trade name of Jewel. In 1876 it introduced the use of nickel-plated stoves. The Michigan Stove Company made almost 200 varieties under the Garland label. In the 1880s Detroit Stove employed about 1,300 men and Michigan about 1,000.

Both companies quickly developed extensive markets for their products and Detroit shipped hundreds of carloads yearly to other countries. Detroit Stove had distribution centers in Stockholm, Frankfort, and London, as well as in Buffalo, St. Paul, and Chicago. Besides heating stoves, both the giant stove companies made full lines of furnaces, domestic and commercial ranges, and cast-iron holloware.

Many skills and trades essential to this dynamic industry were fostered through apprenticeships and other training. Industrial designers, metallurgists, production specialists, model-makers, pattern-makers, core-makers, molders, welders, grinders, sheet-metal workers, riveters, metal-finishers, platers, enamelers, polishers, painters, assemblers, machinists, mechanics, millwrights, tool-and-diemakers, supervisors, foremen, and a host of other specialists flourished.

The fledgling automobile industry thus found in 1900 a ready-made pool of expertise and skills it probably could not have found in any other place having Detroit's other advantages. In addition, the new industry found the executive, management, financial, purchasing, inventory management, and the far-flung sales functions were readily transferable from stoves to cars. In a very real sense the stove industry prepared the way for the automobile industry.

EARLY UNIONIZATION

In another respect, too, the older industry foreshadowed — though by many years — what the motor industry would face. Key elements of the stove industry labor force were unionized almost from the beginning. The industry was one of the first strongholds of organized labor in Detroit, and strikes were not unknown. In fact, Lafayette Crowley, the grandfather of the author of this article, was brought on in 1880 as general manager of the Detroit Stove Works — a position he held, along with a seat on the board of directors, for nearly 30 years — from a large St. Louis foundry, in part because he had successfully negotiated the end of a nine-week molders' strike there.

The two principal companies competed vigorously and prospered handsomely. As a result, a significant part of the start-up capital of the automobile industry came from the profits realized by the investors in the stove industry. The list of officers and directors of the stove companies contains the names of many of Detroit's leading capitalists in the 19th and early 20th centuries, men who were also backers of the new motor cars, including such familiar names as Dwyer, Tefft, Barbour, Henry, McMillan, Ledyard, Palms, Irvine, Mills, DuCharme, Moran, and Campau.

Mention should be made of two smaller companies that played lesser roles in the development of the stove industry in this area: The Peninsular Stove Company, founded in 1881, with its works at the corner of Fort and Eighth Streets, and The Art Stove Company, purchased by Jeremiah Dwyer for his son William, a successful operation for many years until its sale to the Detroit Stove Works. The Peninsular ceased operations in 1931 and the plant was subsequently demolished.

By 1908, Michigan Stove was widely advertising that it operated the "world's largest stove factory." The company was also well-known from its house-sized stove model built for and exhibited at the 1893 Chica-

To promote its products, Michigan Stove Company built this huge wooden replica for the 1893 Chicago Columbian Exposition. Later it stood beside the factory on East Jefferson and then on the State Fair Grounds. (Burton Historical Collection, Detroit Public Library)

go Columbian Exposition. Measuring 23 feet high, 30 feet long and weighing 15 tons, it was an important local and tourist sight for many years. By 1927 it was on a pedestal near the Belle Isle Bridge but was moved to the State Fairgrounds in 1965 and in 1972 turned over to the Detroit Historical Department. The landmark wooden stove is fondly remembered by several generations of Detroiters.

By 1920 stove manufacturing was changing from basically a foundry operation to one involving sheet metal and stamping, as gas came to be the preferred cooking fuel and the design of domestic ranges was altered accordingly. Colorful, attractively-designed ranges intended for modern kitchens made their appearance, and Detroit's production of stoves, ranges, and furnaces averaged 350,000 units annually in the 1920s, enough to maintain Detroit's world leadership by a comfortable margin. Only the automobile industry outranked the stove industry as a processor of metal. The stove industry in Detroit used an average of 25,000 tons of metal annually during peak production years of the 1920s.

AMICABLE UNION

In 1925 the Detroit Stove Works, always somewhat larger and more profitable than the Michigan, took over the latter company in a friendly merger. Operations were consolidated at the East Jefferson-Grand Boulevard site and the East Jefferson-Adair plant was demolished and the site sold. The company then became the Detroit-Michigan Stove Company, easily the world's largest.

Most of the officers and directors of both companies were retained. Profits continued at a high level as the building boom of those years created a ready market for the company's products. In 1929 the common stock was listed for the first time on the Detroit Stock Exchange. Coming out at $9.00 a share, it promptly climbed to $11.00 only to fall to $5.00 in the stock market crash of 1929 and to as low as $.50 a few years later.

A year later, in 1930, it was evident that the stove business was in a depressed state and worse was to follow, in common with the rest of the economy. The company cut its dividend to $.10 a share from $.60, and that was the last one it paid for more than a decade. One unprofitable year followed another, with few exceptions, and layoffs and other drastic economies were the rule. For a few years of the Depression the stock was frozen, with no trading whatever. The Detroit-Michigan Stove Company survived and returned to profitability early in the 1940s, but it never recovered its former leadership position.

World War II brought a complete shutdown in normal production, except for cooking and heating equipment produced for the armed forces. A new facility was acquired on Mound Road and the Metal Fabricating Division was formed to turn out armor plate. That proved very profitable to the company during the war years.

In 1945 a tremendous pent-up demand for stoves existed and the Detroit-Michigan Stove Company was able to move quickly into the market. Its stove manufacturing capability had been mothballed and was intact, ready to go without reconversion. The company was the envy of other Detroit industries that experienced long and difficult postwar conversions and turnarounds. Then followed several years of peak production, with handsomely-styled products for the trade and extra dividends for the stockholders. The stock went on the New York Stock Exchange.

In 1948 the Detroit-Michigan Stove Company took over the A-B Stove Company of Battle Creek, which had well-established lines in the electric range market and in compact units for trailers and mobile homes. Those were markets in which the Detroit-Michigan company had not been represented and they helped to fill out its product line. Eventually the Battle Creek plant was sold and its operations combined with those in Detroit.

DEMAND COOLS OFF

The postwar boom represented the last period of consistent high profitability for the Detroit stove industry. By the early 1950s the accumulated demand had been satisfied and conditions in the American stove industry were increasingly unfavorable to the Detroit-Michigan Stove Company.

As with other Detroit manufacturers of a variety of products, the stove company was a high-cost producer in an industry where there was considerable excess capacity and consequent pressure on selling prices. Much stove manufacturing was being done in the South, in newly-built plants with low labor costs. The Detroit plant was obsolescent and the old advantages of nearness to raw materials and to water transportation no longer applied.

During this difficult period the Metal Fabricating Division was liquidated and the property sold. The company found itself increasingly uncompetitive and profits were hard to generate. In fact, red ink began to appear on the operating statements and the trend was ominous and apparently irreversible. The days of Detroit's venerable stove industry appeared numbered.

Detroit-Michigan Stove Company, however, was still headed by seasoned and resourceful executive officers and directors who did not panic in adversity. A merger was arranged in 1955 with the Welbilt Stove Company, of Maspeth, New York. It was a large, family-owned maker of stoves, ranges, furnaces, and related products, including air conditioners and refrigerators. The addition of the Jewel and Garland lines was a good fit and the New York Stock Exchange listing was attractive.

The merged companies were called the Welbilt Corporation. The Detroit operation was continued for a couple more years, when all production of stoves in Detroit came to an end. Detroit, which had for so long been extremely hospitable to an entire industry — indeed had pioneered it — had made its last stove. Welbilt demolished the East Jefferson plant shortly

This Jewel stove from a 1908 catalogue shows the elaborate iron castings which went into its manufacture, and provided Detroit with a cadre of experienced metal workers for the auto industry. (Donald M. D. Thurber collection.)

after its activity ended, after removing whatever equipment was usable, and the cleared site was sold to Uniroyal as a parking lot. Combined with the now-cleared Uniroyal site, it awaits re-development.

Welbilt, now headquartered in Connecticut, carries on progressively and generally profitably. Its Detroit roots are acknowledged by its use of 1864 as its founding date. Its stock is traded on the NASDAQ Exchange. For those looking for vestiges of a vanished Detroit industry, it is worth noting that Welbilt still markets some of its lines under the Jewel and Garland names.

In fact, the Garland line of heavy-duty commercial cooking equipment is still the world leader in that market, as it has been for over a century, with plants and distribution centers in several foreign countries as well as in the United States. Think of Detroit the next time you are in a fast-food place, and look for Garland — it's probably there. Then, as you drive away from the restaurant in your car, think of the automobile industry and the stove industry that helped to spawn it.

Victory 'Garden'

Manufacturing firms mutate,
mushroom as makers
of munitions, other war materiel

by Hugh Wray McCann

Top: *The "Arsenal of Democracy" cranking out M-3 tanks at the Chrysler Detroit Tank Arsenal, August 3, 1942. (Chrysler Historical Collection)*

On Dec. 29, 1941, almost three weeks after the Japanese attack on Pearl Harbor plunged the United States into World War II, President Franklin D. Roosevelt, in one of his famous "fireside chats," told Americans: "We must become the Arsenal of Democracy."

With that directive, he marshaled the economic might of the nation to an epic display of mass production the world had never known. By the time the war — the greatest in the history of human conflict — was over, American engineers and assembly-line workers had built 124,000 ships, 300,000 war planes, 41 billion rounds of ammunition, 100,000 tanks and armored cars, 2.4 million military trucks and produced 434 million tons of steel and 36 billion yards of cotton textiles.

In tackling this prodigious task, Michigan, then with only four percent of the nation's population, obtained more than 10 percent of the nearly $200 billion awarded in major contracts by the U.S. and foreign governments from June 1940 to September 1945.

Wayne, Washtenaw, Oakland and Macomb counties accounted for more than 70 percent of the state's production. And of Detroit, historian Alan Clive wrote: "No American city ... carried out more war work." Engineers in the metropolitan area made a monumental contribution to Allied victory, more than vindicating Roosevelt's historic exhortation.

Their success traces back to World War I when, even without Roosevelt's ringing exhortation to inspire them, they made their first contribution to an Allied victory in history s first global conflict.

After the war erupted in 1914, Detroit factories took munitions orders from England, France and other European governments. Consequently, the entry of the United States into the conflict in 1917 found several factories ready to step up to a wartime footing. "Practically every plant of any size was swung over to this business," records a City of Detroit historical account. "And in spite of the fact that approximately 60,000 skilled workmen left for the fighting front, the industrial army at home was near 275,000."

NUTS & BOLTS OF POWER

America's entry into the First World War sent into battle Detroit's highly developed system of standardization, high tolerances and interchangeablity — the nuts and bolts of mass-production. The automobile industry, lacking any experience in military manufacturing, unleashed a flood of trucks, ambulances, field-kitchen trailers, artillery subassemblies, tanks, airplanes, airplane engines and submarine chasers. This materiel tipped the war decisively in favor of America's exhausted allies and established the United States as a world power.

Railroad-equipment-builder American Car & Foundry Co. was one of the first Detroit factories to get into defense work. In 1915 it began filling orders for artillery shells from Britain. France had a contract with the Dodge Brothers to build recoil mechanisms for howitzers. Buick was building subassemblies for tanks and tractors for Britain. And Packard was developing the Liberty aircraft engine.

In 1917, when the U.S. became a co-belligerent, American Car went on three shifts a day to make shells. The Ford, Lincoln, Dodge and General Motors plants joined with Packard to mass-produce 8- and 12-cylinder Liberty aircraft engines. Fisher Body built airplane fuselages, and the Ford plant at River Rouge constructed Eagle subchasers for the Navy. GM also turned out 20,000 trench-mortar shells a month while 90 percent of its truck production went off to war. The army adopted the Cadillac V-8 as its standard staff car. And the Buick Model 16AA ambulance became as familiar on European battlefields as the jeep would become two decades later.

Following the armistice in November 1918, the auto industry was demobilized. Although domestic auto production has never ceased, it found a sales backlog of half a million cars and trucks in the postwar marketplace.

Within 15 years war clouds massed ominously on the horizon across the Atlantic. As part of their rearmament programs, European nations began placing orders for military equipment with the United States.

Auto companies were the first to become involved in making products for what would become World War II. They supplied Britain and France with war materiel during the 1930s without interrupting the production of cars and trucks for the domestic market. But in 1941, following the Pearl Harbor attack, they confronted the daunting task of totally reconverting all their facilities to war work.

Within days of the Dec. 7 attack, the government tasked the auto companies with building 75 percent of all aircraft engines, more than one-third of all machine guns, nearly 80 percent of the tanks and tank parts, half of the diesel engines and all of the motorized units for the army.

The reconversion of the auto industry was masterminded by William S. Knudsen, a former president of General Motors Corp.

SIGNIFICANT SITDOWN

Most auto builders date their transformation to a memorable meeting in Detroit on Oct. 25, 1940. Everyone who had anything to do with the manufacture of automobiles attended — from primary producers, parts and appliance fabricators to tool and die makers.

Knudsen outlined principles and practices for farming out subcontracts. And when America entered the war, these principles enormously streamlined the participation of thousands of small industries in the national defense program.

Knudsen announced that he intended to use the auto industry to back up the airplane industry. As a result, DeSoto, Chrysler, Hudson and Goodyear Rubber Co. were soon building a long line of parts for Glenn L. Martin Co. of Baltimore. Murray Corp. did the same for Boeing's B-17 Flying Fortress. And Fisher Body Division of GM backstopped parts-making for North American.

Within months, engineers and designers from GM, Ford and Chrysler and from hundreds of parts and subcontracting firms swarmed through the plants of the aircraft industry. They made rough sketches, took notes and soaked up airplane expertise and know-how.

Consequently, two years after the Pearl Harbor attack the auto industry was turning out:
  • complete B-24, Grumman TBF Avenger and F4F Wildcat aircraft and Waco CG4 gliders;

- parts for the B-17, Martin B-26, North American Mitchell B-25, Curtiss SB2C Hell Diver, Douglas A-20 Boston, Curtiss C-46 Commando, Consolidated C-87 Express, Douglas C-64 Skymaster and Vought OS2U Kingfisher;
- engines for the North American P-51 Mustang, Bell P-39 Airacobra, Curtiss BT-14 Valiant, and for Britain s Avro Lancaster, DeHavilland Mosquito and Vickers-Supermarine Spitfire.

On Jan. 20, 1942, the newly created War Production Board halted production of passenger cars and light trucks. Thereafter, auto engineers spent the war mass-producing products to vanquish the Axis powers. They turned their brains and hands to making arms and ammunition, military vehicles, artillery and airplanes and equipment ranging from gyrocompasses and range finders to top-secret devices for atomic bombs.

The honor roll in this colossal undertaking contains too many examples to list them all. A few citations must stand proxy for the many.

Ford fabricated the biggest airplane plant in the world — Willow Run, which built four-engined B-24 Liberator bombers. A mile long, quarter-mile wide and costing $100 million, the plant was larger than the combined prewar plants of the major airplane manufacturers of the day — Boeing, Douglas and Consolidated. The facility built in excess of 8,500 Liberators — more than one every hour.

Chrysler constructed its tank arsenal in Warren in 10 months and began shipping units by September 1941. The arsenal built more than 25,000 tanks.

Mass production of the Browning .30-caliber machine gun went to GM's Brown-Lipe-Chapin, AC Spark Plug, Frigidaire and Saginaw Steering divisions. In mid-November of 1940, Saginaw produced its first

Tanks, trucks, airplanes, guns were the stuff of Detroit in the wars. One of GM's unique contributions was this "Duck" amphibious cargo truck used in landings from Normandy to Iwo Jima. (General Motors)

model — seven months ahead of schedule. In March, 1942, when the contract called for delivering 280 weapons, Saginaw shipped 28,728 — and dropped the price per copy from $667 to $141.44.

In tackling production of the famous Swedish-designed Oerlikon cannon, Pontiac Division engineers virtually redesigned the entire piece. Their simplified breech casing, cut machine time from 240 hours to 90. The new design reduced total production time by 35 hours and trimmed $166 from the unit cost.

Packard engineers completely redrafted the blueprint for Britain's Rolls-Royce Merlin aircraft engine. They did so in order to achieve the one-ten-thousandths-of-an-inch tolerances demanded by Detroit's mass-producers. Packard delivered the first nine Merlins at a cost of $6.25 million, with the company's "reaping" a profit of $6,206 on the deal.

From Sept. 1, 1939, when the war erupted in Europe, until August 1945, when Japan surrendered, the auto industry delivered almost $50 billion worth of war materiel. Thirty-nine percent of it consisted of aircraft and aircraft parts, about 30 percent was military vehicles and parts, and 13 percent went into tank production. Marine equipment, guns, artillery and ammunition were among the other major items.

Among the unsung heroes of the home front were the Detroit area's many small war plants, those with less than 500 employees. "Without these plants, it would have been impossible for the plane factories, tank arsenals, truck makers and shipyards to achieve their production goals," said M. A. Holmes, regional director of the Smaller War Plants Corp. "The average American thinks of war production in terms of finished bombers, tanks, guns and ships. He forgets that each of these major items contains perhaps hundreds of parts and subassemblies produced by concerns whose names wouldn' t even be recognized."

Brains & Brawn

Detroit brains and production expertise stepped into the breech during lesser wars in Korea (1950-53) and Vietnam (1964-75).

Within weeks of the invasion of South Korea by its northern neighbor, Ford, Buick and Nash were negotiating with the Department of Defense for the production of Pratt & Whitney engines for B-36 Convair bombers, Boeing B-50 medium bombers, and Fairchild C-119 and Douglas C-124 transport aircraft.

In the first year of the war, Detroit received the bulk of $7.6-billion worth of contracts awarded to Michigan. Ford built a plant in Livonia to produce M-48 tanks. Packard constructed a facility in Utica to manufacture jet engines. Among the principal holders of contracts for tactical vehicles were GM (2 1/2-ton trucks), Dodge (ambulances) and Fruehauf (light and heavy trailers).

The $7.6-billion figure represented the peak of military expenditures in Michigan during the Korean war. And it is well worth noting that in this conflict, the government spent 9.5 percent of all its military procurement funds in Michigan — primarily in the metropolitan Detroit area. (The comparable fraction for World War II was 10.5 percent.) Three years later, when the armistice agreement was concluded, the state's defense spending — the bulk of which continued to be concentrated in metropolitan Detroit — fell to approximately $650 million annually.

In mid-1966, when America escalated its military involvement in Vietnam, millions of additional dollars were pumped into the metropolitan area for the purchase of trucks, jeeps, engines and M-60 tanks. That year, the state's defense contracts hit $1 billion, a peak not recorded since the Korean conflict, with Detroit once again getting the vast bulk of the work.

Ford astounded its critics and nay-sayers in two world wars by mass-producing the seeming impossible — Eagle boat subchasers at Rouge in 1918 and these B-24 bombers at Willow Run, photographed June 24, 1943. (Mike Davis collection)

In 1971, as U.S. participation in the Vietnam war was in its decline, Chrysler was working on a $52 million order for M-60 tanks, and GM had a $34 million contract for automobile engines, spare parts, armored personnel carriers, trucks and M-16 rifles.

And in 1991, the unique human and industrial resources of the Detroit metropolitan area went into action once again on the side of the U.S. and its allies. M1A1 Abrams battle tanks built by General Dynamics in the Sterling Heights arsenal outfitted U.S. Army armored spearheads executing Gen. Norman Schwarzkopf's epic left hook around the opposing forces, ending the Gulf War after 100 hours of ground-based hostilities.

As demanding as the production feats recorded during the Korean and Vietnam wars were, the auto industry once again demonstrated its amazing flexibility by delivering the goods without interrupting the flow of vehicles for the nation's civilian market.

Experts say that modern technology has profoundly altered military strategies and doctrine as they would apply to world wars of the future. So it seems unlikely that the auto industry will ever again face a production demand as prodigious as that of World War II. Be that as it may, many recall those momentous years as the City of Detroit's finest hour.

For Michigan and the United States, it was "probably the greatest collective achievement of all time," wrote Donald F. Nelson, chairman of the War Production Board. "No other such ... effort was ever attempted by the human race."

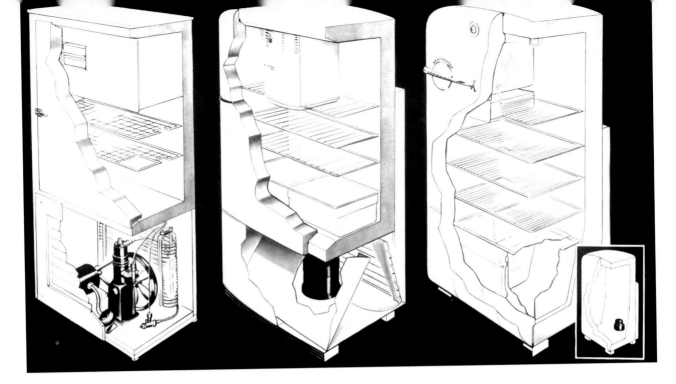

Advances in refrigerator design are shown by this Kelvinator publicity release illustrating 1925, 1940 and 1950 models. (Frank Buttler collection)

A Chill on Appliances

*State fridge makers
nearly frozen out
by auto union pay rates*

by Frank Buttler

Home appliance development and manufacture in Detroit preceded its "Motor City" role, but America's romance with the auto obscured the city's pioneering role in appliance invention and manufacture. For at least the last third of the 19th century and first third of the 20th century, Detroit's industrial reputation was based on production of heating and cooking stoves by at least five manufacturing companies.

Detroit's claim as the birthplace of electric refrigeration for the home came some 50 years after stove manufacturing began.

The first successful household refrigerating system was an idea that became reality, thanks to the founders of Kelvinator who came together in Detroit. There were three men involved in the pioneering work. First was Edmund J. Copeland, a former Buick executive and first president of the company, who in 1914 led the team which overcame early problems leading to a leakproof sealant of the chemical refrigerant and a trouble-free thermostat to maintain safe food-keeping temperatures. A second founder was Nathaniel B. Wales, a Boston, Mass., inventor attracted to Detroit by the young auto industry, who had an idea for mechanical refrigeration. The third was Arnold H. Goss, of Detroit, former secretary of Buick.

Guardian Frigerator Company, forerunner of the Frigidaire electric refrigerator, also was started in Detroit in 1914. Its 1919 acquisition by Will Durant for

General Motors led to its move to Dayton, Ohio, where under Charles Kettering's inventive guidance it became the nucleus of GM's huge home appliance business.

Kelvinator's claim to being first was not disputed although the Frigidaire name was to gain household recognition as virtually a generic name for the appliances. Kelvinator was to remain a hometown company until it was sold in the late '60s. Its still-standing 1927 Plymouth Road headquarters in northwest Detroit was jokingly referred to as the "Tower of the Cathedral of Refrigeration."

Copeland, who after founding Kelvinator went on to head another refrigerator company bearing his own name, later explained the obstacles to early refrigeration development: "After we finished our first compressor, we encountered all the problems which had floored everyone else. There was leakage both in and out on the suction side of the compressors. If we couldn't seal those gases, we were through. We used sulfur dioxide, and any moisture that got in would freeze the compressor up so you had to break it open with a hammer."

Copeland and his foreman experimented with genelite, a porous, powdered compound of oxide and graphite produced by General Electric. Their temperature experiments led to the material's successful application as an effective seal. "For years we had everybody buffaloed. Nobody knew how we made genelite hold gases," Copeland recalled.

The company's first refrigerating mechanisms were built during 1914 in a machine shop on East Jefferson at Chene in Detroit. In 1916, the company incorporated as the Electro-Automatic Refrigerating Company, Inc. Two months later, it was renamed Kelvinator Company in tribute to Lord Kelvin, the British scientist who had pioneered the principles of refrigeration.

During the next 10 years refrigerating mechanisms were remote units installed in existing ice-boxes which had achieved cooling using large blocks of ice. The Kelvinator mechanism was placed at the side of the cabinet or even in the basement because of size and sound. Connections to the cabinets were relatively simple. Millions of usable ice-boxes were converted to the Kelvinator system, so it was not until 1925 that Kelvinator brought out the industry's first self-contained unit — the entire cooling system in one cabinet.

CORPORATE DEVELOPMENT

Public response led Kelvinator to purchase its major cabinet supplier, a Grand Rapids company and to market a second brand name, Leonard, after the cabinet company's founder. The company broadened its prod-

The earliest electric refrigerators by Kelvinator utilized the old "ice box" cabinets and put the compressor mechanism in the basement, as shown in this drawing. (Frank Buttler collection)

uct line by acquiring the Nizer Corporation, then the largest manufacturer of ice cream cabinets. It also created a subsidiary, Refrigeration Discount Corporation, to finance purchases by dealers and their customers. Kelvinator of Canada, Limited, was formed to assemble and distribute refrigerators.

Still later in that decade, the company brought in a strong new president, George W. Mason, who had an impressive managerial record with Chrysler and Copeland Products. Mason guided the company to growth and an expanded product line including the acquisition of Ranco Inc., which had supplied Kelvinator with thermostatic controls. Under Mason's leadership, Nash-Kelvinator Corporation was formed in 1937 by merging with Nash Motors.

As a division of the corporation, Kelvinator expanded its product line to include freezers, electric ranges, water heaters, air conditioners and disposers. At the same time, its commercial line was expanded to include water coolers, beverage dispensers, frozen food display cabinets and similar products. In 1952, the home appliance line was completed with the addition of laundry equipment by the acquisition of Altorfer Bros. Company (ABC brand) of Peoria, IL. Kelvinator became a division of American Motors Corporation when Nash and the Hudson Motor Car Company merged in 1954.

Kelvinator exported its products as early as 1924, only 10 years after the first successful refrigeration systems were built. The first export shipment consisted of a dozen refrigerators made in Detroit and shipped to Shanghai. From that beginning, international operations became worldwide; manufacturing was carried on in subsidiary companies in Canada, England, Italy and Puerto Rico and in licensee companies in 18 plants in 17 additional countries. Distribution of Kelvinator and Leonard appliances was carried on in 141 countries through more than 220 distributors. Kelvinator's name is still seen in some of these countries today as it is in the U.S.

Kelvinator contributed many "firsts" to the home appliance industry from its beginning to modern times, including:

1914 — First successful automatic refrigeration for the home
1925 — First self-contained refrigerator
1935 — First refrigerator with across-the-top frozen food chest
1948 — First top-to-bottom refrigerator
1952 — First automatic-defrost without heating elements
1955 — First side-by-side freezer/refrigerator combination, the Foodarama
1955 — First disposable foil oven liners
1960 — First refrigerator with foamed-in-place insulation

During the frantic post-World War II buying period, Kelvinator scrambled to buy steel to meet the market demand for home appliances. Extreme retail competition lead to the success of mass merchandisers or discount retailers who replaced conventional franchised appliance dealers. As a consequence, manufacturers began to lose control of retail pricing as retailers largely abandoned traditional brand loyalty and brand exclusivity. Instead, they demanded more and more price concessions based on the economies afforded by larger shipments, warehousing, promotional and advertising allowances, and other factors.

Uneven Playing Field

Manufacturers such as Kelvinator were subject to varying labor costs depending on the industry in which the parent company was established. While many appliance manufacturers' plants were organized by unions in the traditionally lower-cost electrical industries, Kelvinator, Frigidaire and Norge were divisions of companies represented by the UAW. Consequently wage levels at their plants were closer to those of the auto industry than to others in the home appliance industry.

Ford Motor Company joined the appliance industry in 1962 by purchasing ailing Philco, manufacturer of both "brown goods," (radios, television sets, etc.) and "white goods," (traditional kitchen appliances), but its Philco employes had been organized by the electrical worker unions. Kelvinator was even more susceptible to non-competitive labor cost pressures because its Grand Rapids plant and its plastics plant at Evart, Michigan, were suppliers to AMC's automotive division.

While Kelvinator's history was marked by acquisitions during much of its early and Nash-Kelvinator periods, the '50s were to reverse those earlier expansions. Fierce competition for survival in the auto industry and the need for cash led American Motors to spin off subsidiaries to pay for design and development of its compact cars under the leadership of George Romney. Supplier companies such as Ranco, known for its Oasis and Ebco names in office water coolers, were sold. The Plymouth Road plant which had continued to produce refrigeration compressors was closed in 1957 and manufacturing consolidated at Grand Rapids. What was once the "Tower of Refrigeration" symbolized AMC world headquarters in the '60s and early '70s.

Marketing conditions which ravaged the home appliance industry beginning in the late '50s took their toll on all appliance companies but the largest such as General Electric and Whirlpool were able to survive. Others became targets for acquisition and merger. Kelvinator fell to this trend when American Motors sold the division in 1967 to Cleveland-based White Consolidated, a conglomerate which would gobble up many brand names including Frigidaire, Westinghouse, Philco and many others. In what proved to be a harbinger for American Motors, ReDisCo, the finance subsidiary, was sold to Chrysler in this same time frame.

Today the Kelvinator name is continued along with many other brands, many produced in Grand Rapids under a single corporate ownership.

Burroughs Adding Machine Co. plant on Second Boulevard, about 1919. (AKA)

Just Doesn't Add Up

Burroughs, others lost out to teen-age tinkerers: go figure

by Jon Lowell

D on't tell me about the computer industry. This is the only business in history where a Fortune 500 company can get blindsided by a 15-year-old tinkering in a garage."

That comment wasn't made in California's Silicon Valley. It was made sardonically in the early 1980s in Detroit in the executive suite of what was then one of the world's largest computer companies, Burroughs Corporation.

Located almost literally in the shadow of the General Motors Building in Detroit's New Center area, Burroughs was battling to remain competitive in a business and technological maelstrom that rages to this day. The fight continues for Unisys Corporation (a 1986 merger of Burroughs and Sperry Corporation) in Blue Bell, Pennsylvania, but for the veterans of Burroughs the warm memories are of Detroit triumphs and technical milestones that went largely unnoticed in a city dominated by the auto industry.

That was just fine with early leaders of the company who were content to run a paternalistic and hugely profitable company that provided employes such perks as a company-owned country club in relative obscurity. Latter-day chairmen such as Ray W. MacDonald and former Treasury Secretary W. Michael Blumenthal were less pleased with the low profile of what had become a $5 billion company.

Computers were a logical evolution of a company whose founder, William Seward Burroughs, invented the world's first practical adding machine (the abacus is another story) in 1886 and revolutionized the lives of bookkeepers and accountants. Started in St. Louis, the company moved to Detroit in 1904 to tap the area's rich supply of skilled craftsmen. Originally called the American Arithmometer Co., it became the Burroughs Adding Machine Co.

VALUE-ADDED

Business machines spun from adding machine technology were, in fact, the computers of their day, allowing companies to grow rapidly and still keep reasonably sophisticated records without rooms full of people filling out ledgers. The company thrived until the onset of the Great Depression. By the early 1930s, to pull itself out of an environment that was at best stagnant, the company had brought in a fresh generation of managers who were casting about for new technical horizons.

World War II opened those horizons. Burroughs quietly became the large scale producer of the super-secret Norden bombsight, credited by historians with revolutionizing aerial warfare and contributing importantly to the Allied victory in Europe.

During the 1940s, Burroughs also built the electronic memory for ENIAC, the world's first electronic automatic computer. These days, that nightmarish roomful of vacuum tubes and miles of hand-wired circuits would be easily outperformed by the personal computers on most desktops.

During the 1950s and 1960s, Burroughs Corporation (as it renamed itself in 1953) scored a number of technical firsts. The company developed the world's first desk-sized computer in 1954 and the first operational transistorized computer systems used in every Mercury and Gemini manned spaceflight mission. Defense systems became an important part of the Burroughs product mix. Sales grew fourfold from $94 million in 1948 to $389 million in 1960.

In 1959, sensing a good thing, RCA made an unsuccessful attempt to buy the company.

Deciding to go it alone, Burroughs stepped up its efforts in its area of traditional strength — business systems. Focusing its efforts on large, main-frame and medium-sized computers, it tried to take on the General Motors of its industry, IBM, virtually across the board.

In some industries, notably banking, hospitals, airline reservations systems and large government computer installations, Burroughs gave IBM fits. The company developed one of the world's primary inter-bank international electronics funds-transferring systems, SWIFT, and developed a series of large-scale computers that became the darling of computer sophisticates. Sales soared to $943 million in 1971 and $2.8 billion by 1979.

BIG, AND BLUE

In the end, however, the biggest problem turned out not to be the trench warfare with IBM's marketing juggernaut. Ironically, both companies got blindsided by those pesky kids in garages who developed microcomputers and software that simply changed the rules of the computer game.

When Burroughs merged with Sperry, it had to shed 24,000 of 121,000 total employees. Survivor Unisys reported revenues of $7.7 billion and employment of only 49,000 at the end of 1993.

Burroughs' former world headquarters these days houses only a fraction of the people who as late as the mid 1980s spilled over into a series of adjoining buildings. And Unisys continues to struggle to find its place in a business world redesigned by technically creative children.

Computers have replaced old-fashioned mechanical adding machines.

Detroit's River

Like the wheels of industry, it's always rolling, whether noticed or not

by Douglas Williams

Top: *The ore ship WILLIAM CLAY FORD being launched in 1953 at the Great Lakes Engineering Works shipyard in River Rouge. (Mike Davis collection)*

The difference between the Detroit River today, and what it was one hundred years ago, is striking. These days — in a sense — the Detroit River "isn't there" for most people in Greater Detroit.

Sure, people see it. Everyone stops and looks occasionally. A passing ship from who knows where. Bright sails stand out; small assorted pleasure and fishing boats are noted; the blue or the ice is remarked upon. The Ambassador Bridge has a certain majesty. Sailboats in the summer are neat. The July 4th fireworks are stunning.

Still, the Detroit River today is static landscape, like a scenery painting — in our minds, it isn't alive — especially compared to anything that might be shutting down a freeway.

The freighters prove this like nothing else. Ninety-nine percent of the people who look at the river daily aren't aware that the ocean ships have the sharp-pointed bows. Or that ocean ships are never more than 700 feet long, because the Welland Locks at the far end of Lake Erie won't take bigger ships.

Most people aren't aware that the lake boats have blunt vertical bows. That the biggest 'lake boats' reach 1,000 feet in length. That 1,000 feet puts them among the very largest moving objects on the face of the earth.

Tell the kids to look for an "apartment" house on the stern. If the stern is flat right across the 105-foot width, it's a thousand-footer. Tell the kids to look for

213

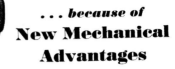

One of several outboard motor manufacturers operating in Detroit was the Caille company, whose product is shown in this 1930 advertisement. Caille's plant was on Second near the Burroughs facility, and the company also manufactured coin-operated devices such as scales. (Mike Davis collection)

the bicycles they ride on the deck of the 600 to 1,000 footers to get from one end to another.

LIVING RIVER; RIVER LIVING

It was incredibly different a hundred years ago. Back then the Detroit River was a vibrant, living thing. It loomed large in everyone's life. It was the lifeline for work, jobs, enterprise. The river decided whether it was a good day or a bad day, whether the year was a good one or bad. What happened on the river affected daily life.

Today — except for some small enterprises, dedicated fisherfolk and sailing types — for most of the people in southeast Michigan, the river could be paved over. Alas.

The metropolitan area is here because the river is. Detroit's first population boom came when the Erie Canal opened in 1825. There weren't any real roads to-and-from here then, just old Indian trails, some "paved" with logs into so-called "plank" roads.

Shipping by water is now, has been, and probably always will be the cheapest way to move goods.

The economic history of the river was written first with furs and the extraordinary European demand for them. Then came lumber and copper. Iron ore and metalworking and automobiles followed.

Today the auto plants look for green fields like a herd of cattle. That's where modern auto plants have migrated.

In 1890 there were 1,236 sailing ships averaging 320 tons working the Great Lakes. Twenty years earlier the number had been near its peak of 1,545 sailing ships averaging 254 tons. That year, 1870, 625 steamers averaging 219 tons were in competition. Cast your mind back to a sailing ship, a working ship powered by sail, carrying 320 tons of goods. Think about how those goods got on a boat, and off. Envision how the ship came into port, and left. Think what it meant for goods and materials to arrive and depart under yards of canvas.

By 1890 there were 1,507 steamers averaging 430 tons. Coming events were foreshadowed under the coal smoke spewing forth from their boilers.

Remember too, it was just a few years over a century ago that the steam-powered ships passed the number of working sailing ships on the Great Lakes. "Sailing is still very competitive in cost," went the common argument. "It takes fewer people to operate and it's not dependent on fuel." Sounds like Roman days, not something dating from when the Engineering Society of Detroit began.

PROLIFERATION & PROFIT

Detroit's river over the past one hundred years, until very recently, was a garden of economic activity. The river was where businesses and industries took seed and flowered, where jobs were, where profits were to be found and livelihoods made. Sometimes those businesses bloomed like they were in a moist, hot jungle.

Often they died and disappeared without a trace.

Sometimes after they died, they were the fertilizer for other huge enterprises that did it all over again.

Consider one example.

Great Lakes Engineering Works was a huge enterprise. In two locations on the river it built large ships through two world wars.

Great Lakes Engineering was the largest ship builder on the Great Lakes, and "equal to any on either coast with two or three exceptions," according to the 1904 Beeson's Marine Directory. At Great Lakes they could "build two to four of the largest type of steel boats at the same time," according to Beeson.

In its late maturity, Great Lakes built the SS William Clay Ford in May 1953. It was 647 feet long, 70 feet wide and drew 35 feet and was called a member of the "Pittsburgh class" because Pittsburgh Steamship ordered three in August 1950.

Today you can look out on the Detroit River from the original wheelhouse of the William Clay Ford. It's been installed, complete, with working radios, at the Dossin Great Lakes Museum on Belle Isle. (The Dossin is a small jewel itself. It was a gift from the Dossin family of soft drink fame, Nehi and Pepsi. On the east side of Belle Isle, it's profoundly neglected but still a real treat).

Take a look at the Museum's model of the Car Ferry Steamer Sainte Marie. She was built by the Detroit Dry Dock Co. in 1893. At 302 feet by 51 feet, with a draught of 24 feet, she was the "largest and heaviest wooden hull operating on the Lakes." Note the propellers at bow and stern.

The last ship built by Great Lakes was the Arthur B. Horner in November 1959, launched two years before the company ended.

In its life, Great Lakes built 328 hulls, 226 of them major vessels. Of those 226, 120 hulls were Great Lakes-faring ships, 106 salt-water ships ("salties" in marine parlance). In addition to those that worked on the water, some worked below it. Great Lakes hulls number 44 to 52 were sections of 3/8 inch nickel-plated steel cylinders, each 23 feet 4 inches in diameter and 21 feet 10 inches long. They were floated to a location in the Detroit River, held by two huge anchors of 22 tons each buried 700 feet upriver, joined into two parallel tubes, and sunk. On June 1, 1910, they began work as a railroad tunnel and function to this day.

The SS Edmund Fitzgerald, in 1995 the most famed lake ship, was built by Great Lakes Engineering Works in Ecorse. Launched June 7, 1958, she was 730 feet long and 75 feet wide, the first of a new class. Some called her the "Queen of the Great Lakes." A glamour ship, she set a record of some kind every year of her short life. On Nov. 10, 1975, with 27,000 tons of taconite filling her three huge cargo compartments, a storm drove her to the bottom of Lake Superior. Today she lives on in Gordon Lightfoot's song, an annual memorial service at the Old Mariners' Church on East Jefferson and the newspaper stories that surface regularly about the ship and her end.

LUXURY ON THE LAKES

Three famed ships called the City of Detroit I, II and III give a hint of how eyes and minds looked at the water years back. The first City carried both passengers and freight around the Lakes for the D&C Line starting in 1878. The last, the City of Detroit III, was launched in 1912 with "rich interiors" and accommodations for 1,440 passengers on seven decks. Seven decks. The largest steel-hulled sidewheel ship of her time at 455 feet long, she worked until 1950. The end of the City III came close to the opening of the interstate highway system, a roadway web that altered the American landscape forever.

The ship business was not just about the big ones.

The street that runs from East Jefferson to the river along the upriver edge of Waterworks Park was known as "Motor Boat Lane" in 1910. Here teemed for many years a variety of boat yards, small shops, engineering works, manufacturing operations, all dedicated to power boating.

Here was born the Belle Isle Boat "Bearcat." At 26 feet and powered by a 200-hp Hal Scott engine, it sold for $2,600. Then came the 26-foot ChrisCraft with a war-surplus Curtiss aircraft engine, selling for $3,700.

Here's where ChrisCraft bloomed, where Gar Wood and Hacker blossomed. All gone today.

Still another example: Gray Marine Motor Co. Few people have heard of it today. A business that was planted in 1906, grew, roared to where it built 20,000 diesel engines in the wartime, pre-freeway days of 1943. Those marine powerplants, from 5 to 225 hp, made Gray the largest marine diesel manufacturer in the world. Today Gray's another unsung industrial ghost.

You would be hard put to find a finer example of this thesis than you'd get in a short talk with Cameron Waterman. Ask him and he'll say "Yes, my dad invented the outboard motor."

Waterman has been a member of the Engineering Society of Detroit since 1959. A registered professional engineer, he retired in 1971 after starting in 1935 at Detroit Edison. Like his father, who had the same name but with a middle initial, B, Waterman was a lifelong Detroiter and Grosse Pointer.

QUICKLY QUALIFIED

Quickly, with the caution of a scientist, Waterman hastens to add that "I can't really say he 'invented' the outboard. But he was the first in production."

Research the topic yourself: it's thin findings.

When Cameron B. was a law student at Yale in 1905, his son tells, he took the Curtiss engine from his motorcycle and hooked it up to the gas light. It ran. Waterman senior had drawings made, then applied for and received a patent, number 851,389. He came back to Detroit and started building and selling small engines that went on the back of your rowboat and made life a whole lot easier.

In the first year, 25 of the air-cooled engines were made and sold. Lake Muskoka, Ontario, proved the change-over point. The cycle-based Waterman outboard ran perfect into the wind but overheated with the wind. Thus a change to water cooling.

Back in Detroit the new powerplant grew popular. Volume from a plant on West Fort in 1907 was 3,000. By 1915 the Waterman had a reversible propeller and removable head. It could also be used for stationary power "around the camp or farm." Also by the end of 1915, more than 30,000 had been sold, very large numbers in those days.

And the end? A market survey in 1917 said there was no future to the business. There simply weren't enough cities close enough to water to interest enough people.

So the business was sold to Arrow Motor and Marine of New York that year. They made the Waterman Porto until 1921; the company went under in 1924.

Waterman wasn't the only Detroit maker of outboards. Caille Brothers made them too, along with "coin operated machines," polite terms for the likes of nickelodeons, penny-for-your-weight scales and slot machines, the last better known by youthful nickel-and-dime gamblers as "one-armed bandits." The office and factory was on Second Boulevard opposite Burroughs Adding Machine. Caille's Liberty Drive Single appeared in 1917 and lasted until 1930. It was a one-cylinder two-hp outboard. The Caille Red Head Model 25 was a two-cylinder, 15-hp, 5-speed outboard with an adjustable pitch propeller. Outboard output ended about 1935.

Other Detroit outboard companies included American, Clarke Troller, Continental, Cross Radial, Sea Gull, Detroiter, Emmons, Gray Gearless, Strelinger, and Sweet.

Now the Waterman outboard company wasn't Great Lakes Engineering Works, the Detroit Stove Works, or any of the industrial operations that took root on the river; boomed and later disappeared. But as a metaphor for all that, the Waterman outboard engine will do.

Forgetting is effortless. We live in a mutating video interstate freeway world. Everything passes the windows at 60-plus miles an hour. Even remembering how central the Detroit River was to everything in this corner of southeast Michigan takes real effort.

Take three points:

• Cadillac built his fort between two rivers, the Detroit and the Savoyard. The Savoyard's gone, the fort's gone, Detroit's still here.

• In the 1920s if you stood in any of Detroit's three huge, famed downtown department stores, J.L. Hudson's, Crowley's or Kern's, you might hear a reference to "wetlegs." Everyone knew it meant Canadians. They came shopping by ferry.

• Waterfront property was crucial to French settlers here. It was their opening to the rest of the world, their interstate of the day.

Thus hundreds of farms up and down all the waterways were a hundred yards wide, called arpent. They were also a mile deep inland. We pay for that to this day. North-south travel in Detroit, along old property lines, is a snap. East-west travel in Detroit is still clumsy, 250 years later.

Drug Industry: The Legit One

Two major medicine makers gave market a shot in the arm

by Bob Cosgrove

The past 100 years have seen a revolutionary change in the manufacture and distribution of fine pharmaceuticals throughout the United States and the world. Detroit was at the forefront of this industry for the first 85 years of the last hundred.

By 1895 our city in a short 40-year period had risen to world leadership in the manufacture of prescription drugs, dental and veterinary products. The two largest U.S. pharmaceutical manufacturers had their headquarters and main manufacturing plants here. They were Frederick Stearns & Company, founded in 1856 and Parke, Davis Company, which traced its beginnings to 1862.

By 1929 Detroit's pharmaceutical industry had over 4,000 employees, a very high percentage of whom were women. Of these, 2,500 were with Parke, Davis (the firm had over 4,100 worldwide), 500 at Stearns, 250 at Nelson, Baker & Company and 200 at the Charles Wright Company. All these firms are gone today as Detroit entities, although several original Parke, Davis facilities survive in the state.

Parke, Davis was a pioneer in setting standards for the pharmaceutical industry. In 1901 Parke, Davis established the first systematic method of technically proving the value of a new medicinal agent before marketing. This was followed in 1902 by the first industrial building specifically erected in the U.S. by any commercial institution solely for the purpose of scientific research. In 1907 they created the first department for field testing experimental medicines, a field which is known today as clinical investigation.

Under the supervision of scientific director Dr. E. M. Houghton, P-D first produced standardized medicines using chemical assay, was given U.S. Biological License Number 1, first produced diphtheria vaccine, sent expeditions throughout the world seeking medici-

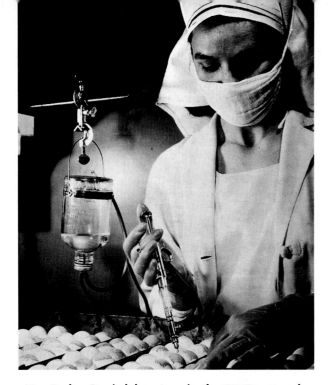

At a Parke, Davis laboratory in the 1950s, a worker in sterilized clothing inoculates incubated eggs with live influenza virus in an effort to develop a successful vaccine. (Detroit Free Press)

nal plants, introduced the mass-produced gelatin capsule, established a worldwide network of manufacturing branches (by 1955 there were over 60), was one of only two U.S. companies providing a complete line of ethical and non-prescription drugs to pharmacists, had the first industrial pharmaceutical library, and published medical journals for physicians as well as technical magazines for pharmacists.

An important contribution, although not unique to Parke, Davis, was the mechanization of production. In the 1890s Parke, Davis had rooms of women hand rolling pills. Just a few years later, Colton tableting machines made to Parke, Davis specifications produced 225,000 tablets an hour. From the earliest days Parke, Davis emphasized quality control, beginning with raw materials and packaging and continuing throughout the manufacturing process, as well as testing the end product even after distribution.

Parke, Davis began in 1862 with Detroit physician and pharmacist Dr. Samuel Pearce Duffield (1833-1916). His manufacturing venture became the partnership of Duffield, Parke & Company in 1866. By 1871 after Dr. Duffield had withdrawn, the firm acquired what would become the most famous corporate name in medicine of the 20th century, Parke, Davis & Company.

DRUG DIVIDENDS

Its principals combined the business talents of the tall scholarly financier Hervey Coke Parke (1827-1899) with the marketing genius of wholesale drug salesman George S. Davis (1845-1930). Parke and

Davis incorporated the venture in 1875, but it remained unprofitable — as it had been since Dr. Duffield's founding — until 1877. The first dividend was paid in 1878. Thereafter Parke, Davis never failed to pay a dividend until its separate identity vanished in 1970.

Under George Davis' instigation, the firm sent researchers to remote parts of the world such as the Andes and the Fiji Islands. Their mission was to find exotic natural materials, particularly plants, from which medicinal products could be extracted. So successful was this operation that a New York City branch was established to import crude drugs as well as export domestic finished product. The crude raw materials were also sold to other pharmaceutical houses.

George Davis assembled an associated team of leading scientists engaged in pharmaceutical research. Some worked directly for Parke, Davis in their laboratories, others at universities where the firm financially supported their research. Great results were produced.

For instance, it was Davis who brought the Japanese born, University of Glasgow educated chemist Dr. Jokichi Takamine (1854-1922) into association with the firm. At the turn of the century Takamine isolated the blood-pressure-raising hormone epinephrine from animal adrenal glands. This product is so well known today that few people realize that "Adrenaline" is a Parke, Davis brand name. It was the first hormone obtained in pure form and the first whose chemical structure was determined.

Another example of Parke, Davis-supported research is Dr. E. A. Doisy of St. Louis, who developed Dilantin in 1938 for the control of epilepsy. In 1946 Dr. George Rieveschl of the University of Cincinnati developed Benadryl, one of the first and most essential antihistamines. The broad spectrum antibiotic Chloromycetin (chloramphenicol) was developed by company research teams in 1949.

Despite all this, Parke, Davis eventually succumbed to a combination of too few new products and the firm's high cost of a union labor force in an antiquated Detroit plant. The final blow was Parke, Davis' inability to replace the profits lost when key patents expired.

Despite P-D's many strengths — including its broad product line, worldwide distribution system, research division and modern outstate plants — the stockholders, controlled by prominent Detroiters, sought more profitable investments. The result was that in 1970, Parke, Davis was merged into the Warner-Lambert Company headquartered in Morris Plains, New Jersey. Today, the Parke, Davis name continues as a Warner-Lambert brand. Its Detroit manufacturing complex was closed in 1980 and sold to the Stroh Brewing Company.

Stroh has redeveloped the 1873 red brick complex. The former administration building on the river is headquarters to Talon Industries. The River Place Hotel is housed in the 1902 research building. The Stroh Place office building, formerly the 1929 concrete-reinforced main production plant, includes a well-known restaurant. The older timber-framed buildings dating back to the 1870s have been made into fashionable apartments.

An Excess of Eggs

The drug firm's Parkedale Biological Farm near Rochester still produces biological products such as tetanus, polio, diphtheria, influenza and other vaccines. Created in 1908, Parkedale once occupied 700 acres to grow medicinal and animal products. The Parke, Davis Holland, Mich., chemical plant is another part of the Warner-Lambert conglomerate. The scope of this operation is staggering. For example, 4 million eggs are used annually in the production of vaccines.

The company's research laboratories in Ann Arbor near the University of Michigan's North Campus opened in 1959 and have been expanded several times. These modern laboratories, and the resource their staff represents, were a major reason for Warner-Lambert's acquisition of Parke, Davis.

Frederick Stearns & Company, which in 1895 was Detroit's other world-renowned pharmaceutical firm, was founded in 1855 as a small one-room laboratory behind the drug store of pharmacist Frederick Stearns (1831-1907). During the Civil War, the Stearns firm gained enormous stature as a purveyor of medicines to Michigan troops. At that time the firm quickly grew into a large four-floor plant in downtown Detroit equipped with steam power, milling and extraction apparatus.

Frederick Stearns was concerned with quackery and in 1876 introduced a concept of non-secret family medicines. He offered a few simple preparations with full usage descriptions and listed their ingredients. Other druggists took to the idea and had Stearns manufacture similar products for them to be sold under their label. This "ethical specialties" system of manufacture introduced by Stearns continues today in the marketing of generic and private brand drugs. By 1895 the firm claimed to operate the oldest and largest laboratory of its kind in the world. Among its many achievements, Stearns introduced the first "house organ" newsletter for pharmacists in 1879.

In 1887 Stearns turned the business over to his sons Frederick K. and William L. Stearns. While the firm continued for another 50 years, it appears to have lost its cutting edge in fine pharmaceuticals when the founder retired. The Stearns company lost its identity in 1937 when it was sold by the heirs to the United Drug Company, whose Rexall drug stores proliferated throughout the U.S. Unlike Parke, Davis in its later years, Stearns products were more of the over-the-counter variety than fine pharmaceuticals. They included the popular mouthwash "Astringosol" and the "Day Dream" line of toiletries.

Frederick Stearns & Company's last office and plant location were the works of architects William B. Stratton and Albert Kahn, beginning in 1900. They survive today at the corner of East Jefferson and Bellevue, just west of the Belle Isle Bridge. Vacant for decades, they were converted several years ago into the fashionable "Lofts" apartments.

Capsule Development

In 1895 gelatin capsules were a major Detroit pharmaceutical product. Developed and manufactured here, capsules, convenient and important in masking taste and easing "swallow," were originally made by hand. The first machine-made gelatin capsules were developed in 1874 by Detroit pharmacist and inventor F. A. Hubel. Introduced in 1876, his factory's production was always sold in its entirety to Parke, Davis. The latter used the capsules in their own manufacturing and sold empty capsules to other pharmaceutical houses. In 1901 Hubel sold his capsule manufacturing plant to Parke, Davis, which remains one of the world's leading producers under the Warner-Lambert banner.

In the 1930s Detroit socialite and inventor Robert P. Scherer developed the revolutionary one-piece soft gelatin capsule machine. Unlike the Parke, Davis capsule, these are filled during the capsule manufacturing process. For many years the R.P. Scherer Company limited itself to filling soft gelatin capsules on contract for firms such as Parke, Davis with medicinals supplied by the customer. Vitamin capsules were one of the early uses of this process. Today, the R.P. Scherer Company is the leading Detroit pharmaceutical manufacturer, although it may be considered more a packager than a research-oriented drug producer.

Another present day Detroit manufacturer akin to and supporting the pharmaceutical industry is Difco Laboratories. Difco is renown for its production of biological media, reagents and the kits. Its products are used in scientific and pharmaceutical laboratories throughout the world.

Working for The Railroads

Area rail engine & car builders had heyday while sun shined

by Bob Cosgrove

By the 1980s, the latest state-of-the-art in locomotive design was this GM Electro-Motive Division SD50, claimed to be the most fuel-efficient locomotive in America. (General Motors)

Detroit's early 20th century renown as a railroad-car-building center began in 1853, when Dr. George B. Russel (1816-1903) opened one of the first commercial rail car shops in the Midwest. This was one of the city's earliest manufacturing industries. In 1859, in part to support his car manufacture, Russel built the first blast furnace west of Pittsburgh near the present Belle Isle Bridge.

The foundry portion of the firm became the Russel Wheel & Foundry, which continued into the 1920s, producing rail car wheels, logging cars, and cast and forged car parts. The Griffin Wheel Company, which moved to Detroit in 1870 from Rochester, N.Y., was another early Detroit rail foundry. Long since relocated to Chicago, Griffin today is the largest producer of rail car wheels.

In 1871 George M. Pullman (1831-1897) purchased the car building part of Dr. Russel's business, the Detroit Car & Manufacturing Company. Until then Pullman had his sleeping cars built to order. For ten years Detroit was the manufacturing and repair hub for the ornate mansions-on-wheels operated by his Pullman Palace Car Company.

In 1881 Pullman built his social-experiment factory town of Pullman, Ill., near Chicago, the largest car manufactory in the world. This was after Detroit's Common Council denied Pullman's request to vacate land to double the size of his Detroit plant. The Pullman Car Works-Detroit continued as a branch until 1893, when it was closed as the market for sleeping cars became increasingly saturated.

The Detroit car builders — including Michigan Car, Peninsular Car and Pullman along with the car-building shops of the Michigan Central and Grand Trunk railroads — fostered a local supplier base including local foundries and the stove companies. The parts supplied included forgings and castings, especially wheels, axles, and car hardware which were also sold to railroads and out-of-town rail car companies. The availability of quality Michigan lumber in the pre-1900 era of wooden cars was a key factor in the rapid growth of Detroit's car-building business.

James McMillan, John S. Newberry and others formed the Michigan Car Company in 1864. Colonel Frank J. Hecker and Charles Lang Freer, with financial backing from General Russell A. Alger and James F. Joy, founded the Peninsular Car Company in 1885. In 1892 these two companies merged to form the mighty Michigan Peninsular Car Company.

In 1895, before the U.S. automobile industry was born, the Michigan Car Company's shops occupied 39 acres at Michigan Avenue and Clark Street near the later General Motors Cadillac Division complex. The Peninsular plant was sited on 52 acres at Ferry and Russell where Detroit's incinerator is now.

In 1898 Michigan Peninsular Car became a cornerstone of a new 17-member car-building cartel, American Car & Foundry, headquartered in New York City. In 1907 ACF employed over 9,000 in its two Detroit plants, turning out 100 rail cars per day. But this was down to 2,500 employes by 1919 and production had virtually ceased by 1930. The older rail facilities and high wage floor dictated by Detroit's automobile plants were undoubtedly factors. In the 1930s and '40s ACF built its line of motor buses in part of the immense former Peninsular plant.

BACK IN BUSINESS

Detroit would again produce rail cars. In the 1950s, Evans Products in suburban Plymouth built cars for hauling automobiles and auto parts as did the old-line structural steel supplier Whitehead & Kales in River Rouge beginning in the 1960s and Paragon Steel in Novi in the 1970s.

Among the cars these firms produced were the 85-foot-long bi-level and tri-level auto rack flat cars introduced in the early 1960s and still the primary means of transporting new automobiles. Before then automobiles were shipped in difficult-to-load box cars. By the mid-1980s all of the relatively small Detroit rail car builders were out of production in a business dominated today by just a few extremely large producers. The former

In 1934, General Motors introduced the first successful diesel-powered train in the country, a streamliner on the Burlington line out of Chicago. By the 1960s, diesel locomotives made museum attractions out of the steam locomotives which had ruled the rails for 130 years. (General Motors)

Whitehead & Kales plants now house car-repair companies doing contact work for the railroads.

Detroit's many diverse manufacturers have long supplied the railroad industry, although in recent years only a few have been dedicated to that market. The Detroit Lubricator Company and the Detroit Seamless Tubes Company supplied steam locomotive components until the last U. S. railroad operating steamers, Detroit's Grand Trunk Western Railroad, "dropped its fires" in 1960.

Today, one of the best remembered African-American inventors is Detroit's Elijah McCoy (1843-1929), who held over 40 patents for steam locomotive automatic lubricators. His name is said to be the source of the expression "The Real McCoy," used to advertise the products of his Elijah McCoy Manufacturing Company.

Great Toys for Girls, Boys

Locally, Daisy blossomed and Lionel chugged along

by Tom Yates

It was the object of dreams for virtually every 10-year-old boy from the 1930s to the '60s. We'd be reminded of it every time we opened or finished a comic book.

There, on the inside or back cover, was the advertisement that inspired our dreams — a slick, four-color pitch for the Daisy "Red Ryder" lever-action saddle carbine, a BB-gun. Its blued steel barrel and receiver almost leapt from the pages. The genuine wood stock glowed with the depth of the finish. You knew it was the real thing because Red's picture was right there beside it. Red was a popular folk figure of the times, immortalized in comic-book fiction.

We dreamed of owning a Daisy Red Ryder, but few of us ever attained it. In tens of thousands of homes throughout the U.S., the admonition rang out: "You'll shoot someone's eye out with that!" We aspired but usually failed.

The Daisy Manufacturing Company, long a mainstay of Plymouth, Mich., has not always been known for its BB-gun production. When the company first started in 1882, it produced windmills under the uninspiring name of the Plymouth Iron Windmill Company. It was 1888 when Clarence Hamilton created an air rifle constructed entirely of metal and presented it to the company's Board of Directors as a potential new product. Lewis Hough, the former vice president and general manager, looked at the prototype and exclaimed, "Boy, that's a daisy!" and a famous name was born.

However, the company's name was not changed until 1895 when it was decided the Plymouth Iron Windmill Company did little to inspire the purchase of an air rifle.

The Daisy Company's worldwide market presence is due largely to Charles Bennett, who dedicated most of his life to the company. He introduced the Daisy air rifle to almost every civilized country in the world except Russia. The Daisy enjoyed almost universal acceptance. Chinese Imperial officials felt the BB-gun was a lethal weapon and refused to allow its import until Bennett, ever the promoter, invited a dignitary to plink away with him. Daisy became so successful in international markets that air rifles have been manufactured in Europe and Japan which were little more than Daisy clones.

LACK OF VISION

If Daisy's executives were successful in the toy field they showed less foresight in other areas. In 1903 Bennett tried to get the Daisy Company to invest in the infant Ford Motor Company. Unfortunately, company attorneys ruled that it was against their corporate bylaws, so the idea was abandoned. By 1912 their relatively small investment would have been worth millions. The directors no doubt felt sincere regret for the legal position. However, Bennett himself personally invested $5,000 in Ford Motor Company, selling out in 1907 — before the debut of the Model T — for what he thought was a huge, five-fold profit. (His $5,000 gave him five percent of the company's stock!)

Daisy was no stranger to innovation. They produced a metal rifle in a time when air rifles were most-

ly made of wood. They were also responsible for the invention of the Daisy Pump Gun, the plunger-type water pistol, a metal "bluing" method still in use by real gun manufacturers and a process that provides a rust-proof finish on parts in various colors.

In all the years of its operation, Daisy has kept out of the firearm or lethal weapon field — mother's admonitions not withstanding — and stayed solidly in the toy arena. Their one constant has been dedication to high quality. As an example, when it was found that Minneapolis youths were salvaging steel ball bearings to shoot in their BB-guns, the company encouraged the use of steel shot, as opposed to the traditional lead shot, and sought to design air rifles to accommodate the safer steel pellets.

Their dedication has not been limited to quality, either. In June 1942, the company halted production of air rifles and began producing dies and other war materials during World War II.

Over the years kids have been enthralled and entertained by Daisy models such as the Buzz Baron, the Buck Jones, the 50th Anniversary Golden Eagle, the futuristic Buck Rogers model, the Rocket Pistol, the Disintegrator, the Liquid Helium Water Pistol, the "Bull's Eye," the Davy Crockett, the Zorro, the Annie Oakley (for girls!) and the current line of "TV Guns of the West."

Daisy manufacturing spent its first 76 years of operation in Plymouth. On May 5, 1958, the company closed down the Michigan operation and moved to Rogers, Ark, for larger facilities (and probably lower costs). According to a Daisy spokesman, the company has built millions of air rifles. They're still in Rogers, and still producing the "All American" BB-gun.

The Red Ryder model is still sold today, still in blue steel with a real wood stock. The retail price of $24.95 is not so far removed from the pre-WWII sticker of $2.95. And I finally achieved my dream of ownership!

RAILROADS IN REPLICA

While thousands of boys aspired to the Daisy air rifle and mothers despaired, many youngsters were more successful in a second, less dangerous quest. If we couldn't have a BB-gun we could at least enjoy the pleasures of an electric train. The preferred selection was a Lionel. Today boys and girls alike, not to mention their parents, can still aspire to own a Lionel train, thanks to Detroit entrepreneur Richard Kughn.

The Lionel company started in New York when Joshua Lionel Cowen invented a small battery-powered electric car that ran on a circular track. Original-

ly it was designed as an attention-getter for merchandise in store windows. To Cowen's surprise, people were more interested in the electric car than the merchandise in the window. Cowen changed his marketing target from merchants to children and changed the world of toys.

Over the years the Lionel trains fascinated small boys and girls and not a few fathers. Throughout its history the company has made a number of moves, bought out competitors and been bought and sold by a number of investors and larger companies. Somehow it has always managed to keep turning out high-quality, realistic-appearing models of full-sized trains.

In 1970 Lionel production was moved from the East Coast to Mt. Clemens, Michigan shortly after being purchased by General Mills. It became part of the food company's Fundimensions toy division. Curiously, the Detroit area had become a fractional model mecca through the manufacture of promotional scale models for the automobile companies.

The operation prospered and grew until 1983 when General Mills consolidated all its toy manufacturing operations and moved them to Mexico. Two years later Lionel trains became a division of Kenner-Parker Toys, a spinoff of General Mills' non-food operations. At that time, toy train manufacturing operations were moved back to Mt. Clemens, but the Lionel future seemed threatened.

In 1986 the business was saved — at least it was so believed by toy train buffs — when shopping center magnate and car collector Richard Kughn bought Lionel and formed the current corporation, Lionel Trains, Inc. One of his conditions was that the company and its manufacturing would stay in Mt. Clemens. He also bought up spare parts inventories from around the country and consolidated them in Detroit.

Under Kughn's leadership, Lionel has prospered far beyond his modest hobby expectations. It has lines of scale model electric trains — especially the solid "O-gauge" — sold both as reasonably-priced toys and as five-figure collectibles: handsome reproductions of 1930s Lionels.

Today Lionel continues to entertain children, parents, collectors and even museum goers. The Detroit Historical Museum has an extensive, working toy train exhibit, a perennial favorite of museum visitors, while Henry Ford Museum usually mounts an electric toy train display at Christmastime. Indeed, where would any self-respecting department store or mall be in December without scale model electric trains?

V. Future Stock

*Technological frontiers offer dividends
if we can master, assimilate developments*

Achieving Continuous Improvement Through The Power Of Suggestions.

All people shown in this ad are Dana people. They are identified and years of Dana service indicated: Photo above (l to r) Gene Chadwell, 18 years; Lisa Stoller, 7 years; Steve Kouts, 29 years. Photo below (clockwise from top) Roy Kincaid, 26 years; Santiago Marillo, 6 years; George Eldridge, 30 years; Andrew Szadkowski, 6 years.

Every day, Dana is working to find a better way for our customers and shareholders. And we do it by tapping into our most powerful resource—the minds of Dana people.

In fact, we encourage every Dana person to submit at least two suggestions for improvement each month. Then we work to implement 80% of those suggestions.

The result is continuous improvement—innovative ideas, better products, greater productivity, and more responsive service.

From Spicer® drivetrain components and systems, to Parish® structural components, Victor Reinz™ gaskets, Perfect Circle® engine and chassis parts, and Wix® filters, we're constantly improving components for today's vehicles and equipment. Dana Corporation Toledo, Ohio 43697.

◆ DANA

People Finding A Better Way ®

One of the more accurate visions of the future was this prospect of a superhighway intersection in 1960 shown in a General Motors exhibit at the 1939 World's Fair in New York. (General Motors)

Technological Future

*Panelist speculate
on this centennial
as a midpoint to the next*

In March of 1995, Mike Davis, editor-in-chief of The Technology Century, moderated a panel discussion on the future with three distinguished faculty members at the University of Michigan. A transcription follows of the wide ranging observations of Dr. David E. Cole, Director of the Office for the Study of Automotive Transportation at the Transportation Research Institute; Dr. Craig Marks, co-director of the Joel D. Tauber Manufacturing Institute, and Dr. William R. Martin, Dean of Academic Affairs and Professor of Nuclear Engineering at the College of Engineering.

Davis: Put yourself back mentally to 1895, with the knowledge you might have had then, holding a similar kind of position. What might you have prognosticated? — and apply it to today.

Cole: One way to look at it is in terms of the macro issues that are significant and are going to influence, in one form or another, our future. An example is globalization. We look for globalization to have really a profound role in our future. What does that mean?

Davis: I think of that more as a corporate, organizational issue than a technological issue.

Cole: No. One of the characteristics of the future is, that it's going to be more and more difficult to separate such considerations. Pure technology can't be thought of outside the context of globalization. One example — a few years ago the question was, who is leading in various areas of technology around the world. Now it's getting to the point where that question

227

is irrelevant because technology moves across boundaries so quickly, in multinational companies, it's a global commodity.

Marks: If you look at the manufacturing world, it used to be that you thought of a manufacturing company here with its headquarters, an engineering department, a marketing department and a manufacturing facility. In the globalization sense, today you may have idea generation in one part of the world, the development activity in some other part of the world, manufacture taking place in several parts of the world, and final assembly being put in still other parts of the world. The ability of teams of people to work together, through improvements in verbal communication as well as information and data transfer in the next few years, is going to revolutionize the way we conduct businesses like manufacturing.

Davis: Again, that doesn't sound technological...

Marks: Those operating characteristics are all enabled by technology.

HUMAN FACTOR

Davis: Yes, but we're talking about a change in human nature that is breaking down some barriers, and what I'm wondering is whether in the technological community these barriers are falling faster than they are, perhaps, in the political community?

Cole: Well, I'm not sure that necessarily the barriers are falling. I think what we're seeing is technology exploding and our ability to interface with it is not keeping up.

Marks: Information technology actually is moving faster than we can use it. It presents opportunities for people who can see and exploit the technology already there. It is just waiting to be used.

Cole: The fields of opportunity are huge for just using what we already have, let alone trying to push new knowledge forward. In fact, globalization and information technology specifically are tremendously important.

Davis: So that's another of your macro issues.

Marks: Most data today is computer-generated. It's a digital world coming about. The ability to store, retrieve and transmit data, and turn that data into information is undergoing a revolution. It's all enabled by digital electronics. Then, the communication aspect — which is a separate issue — enables people, through both audio and visual communication, to essentially eliminate the barriers of distance and time.

Davis: What are some of the other macro issues?

Cole: I think the issue of human resources is important. How are we going to find the people who

have the skills to do what needs to be done? Another issue is the whole environmental/natural resource issue. We need to make sure we have a survivable planet. On the other hand, how do we use the resources that will be required in the most efficient and effective way?

Marks: The new buzzword is sustainable technology.

Martin: International trade barriers, accords, how nations work with each other, all are important for both the globalization and environmental issues.

Cole: Yes, the environment is really a global issue. The trade issues and the trade barriers are another aspect. Increasingly, I don't think we can separate things into packets. One of the things that historically we have not been very good at is systems thinking. Almost as a matter of survival, we have to think in terms of systems.

Marks: Systems thinking is a macro issue in two senses of the word. From an engineer's standpoint, systems engineering requires a macro view of problems. It used to be, you could spend your whole career dealing with a few components in the system, and increasingly that is just not possible. The development of every piece of a complex system is so dependent on the rest of the components that systems engineering is becoming a way of a life.

Davis: Let me play a devil's advocate role here. I know that back in the 60s Ford imposed a systems engineering discipline on the product development area... So what you're saying is that things that were talked about 25 or 30 years ago, now people are beginning to understand and put their arms around?

SYSTEMS ENGINEERING

Marks: In the auto industry, broader recognition of the need for systems engineering has to do a lot with the supplier rationalization issue — which says, as an OE manufacturer I can no longer deal with 3,000 suppliers. I've got to deal with a smaller number of suppliers, because I want them to be an integral part of my development process from Day One, and to take on design and development responsibility for significant sub-systems.

Davis: I guess this what is behind Prince no longer being just a mirror vendor? They have empowered him because he is a good vendor?

Marks: If he takes on systems responsibility, he is able to add vanity mirrors to visors and headliners, add side door panels and next instrument panels and eventually, maybe, design an entire interior system. Sourcing major sub-systems is the only way an OE can get from 3,000 suppliers down to 200, which is where they are headed.

Cole: The complexity of just managing things that are too big is overwhelming. Everyone is trying to simplify the best they can. Consequently we take large tasks or challenges and break them up into smaller parts. We do a great job of dealing with each part, but often a poor job of assembling them back into the whole. This notion applies to much of what we deal with as engineers. We break things into chimneys; separate departments in companies, different disciplines within academic departments, manufacturing and product engineering, or manufacturers and suppliers with, all to often, little communication across the boundaries. We strive to become experts within our particular chimney. Systems thinking requires us to create linkages between these separate domains. In fact those that excel in crossing the interfaces, in bringing different disciplines together, are critical to improving the effectiveness of our profession.

Marks: You're not replacing excellence in individual areas but you are adding to it the capability to design more predictable systems.

Cole: We have not been good at managing interfaces.

Davis: Of course, we are talking, really, about that which we are most familiar with, which is the auto industry. So is it happening elsewhere?

Marks: Systems engineering is increasingly a way of life for engineers. It has to do with the increasing complexity of products, the trend toward sophisticated electronic controls, and our ability to deal with these trends.

Davis: This implies very large organizations. There is not much room in here for the little guy, the small companies, the small suppliers.

Cole: There are great opportunities for people or organizations that are good at managing interfaces. I look at my own background in mechanical engineering, with thermodynamics, fluid mechanics, machine design, etc.; there was very little connectivity between the disciplines. Any product you can buy in a hardware store brings together most engineering disciplines in some way, shape or form. But we are not good at weaving together. The Japanese, however, have brought a culture that appreciates the importance of a more holistic philosophy. Consider a Japanese meal, there is food value but also an attractive and even artistic presentation. Philosophically it is not a large step to a production system such as Toyota's, where there is tremendous holistic thinking — where, for example, you wouldn't think of doing a product design without working with the manufacturing people.

Davis: Well again, this is something I know was talked about, and evidently just really didn't happen, 25 or 30 years ago.

FUEL FIZZLE

Cole: Think back to the 50s, everybody was looking forward to fuel injection — which was the new fuel management system on the block — many were saying that in five years fuel injection would replace carburetion. Several things happened. Fuel injection didn't turn out to be quite as good as its sizzle, and the carburetor didn't stand still. The carburetor became a moving target and there wasn't the enabling technology for fuel injection, in this case electronics, to permit development of an effective injection system. But by the 80s the carburetor had become too complex and couldn't meet the challenge of emissions, fuel economy and driveability — and solid-state electronics was integrated into fuel injection. The carburetor died swiftly.

Marks: And what also triggered that transition was the regulatory requirement laid down. We couldn't handle it any other way. We went to the three-way catalyst and the precise air/fuel ratio control it required. We tried for a few years to achieve that control with carburetors and we just couldn't make it.

I would like to mention another macro issue dealing with smaller companies and globalization. An organization known as Michigan Future has articulated an interesting vision of a new economy which is based, not on a few large companies, but upon 18,000 small manufacturers who are really the backbone of manufacturing in Michigan. What they are saying is, if these firms — through adopting the technologies and management techniques that are now available — are linked together with the communication and information technology we've talked about, they could become not just suppliers to the Big Three automobile manufacturers, but world-class suppliers to anyone in the world.

Cole: They're on a data base that is accessible.

Marks: You put together the right combination of those small companies to do any particular job. No, any one company cannot do the job but today you can conceive of what they call the virtual corporation which is a kind of business arrangement which links these companies together to perform a certain task. It may be a task that lasts for just six months or it may last for six years, and they may be involved in several such linkages. But the flexibility and capability that now exists — in terms of what technology brings and makes possible in those types of businesses — makes that a possible scenario. Are we going to do that? That depends on a lot of things. It depends not only on the technology, which for the most part is not the limiting factor, it depends on people's attitudes and understanding of this, and the way in which we deploy our educational resources and train people, and where we put our efforts. That's what Michigan Future is all about. It is getting that vision exposed to people and figuring out ways to begin to move towards that way of doing business.

Davis: Would this require any changes in the legal system or changes in human nature?

Marks: Not necessarily. It may require some training and education, but mainly it requires an acceptance of the concept — that it is a good way to do business.

Martin: What is interesting is that a lot of companies are driving this — it is not the government that is particularly pushing this.

Marks: It really could mean the difference between the demise and the survival of these companies.

Davis: A natural evolution?

Marks: Because without some kind of operation such as this, many of these companies will not be competitive in the global market.

FLUIDITY

Cole: The real keys to survival have shifted from things like stable, large organizations to quick, smart organizations — that's agility. The entire concept of agility requires new thinking and enabling technology. Whether we're talking about agile manufacturing or the virtual organization means that we truly adopt "form-follows-function." This is a phrase we have used for a long time, but historically we have not done form-follows-function because we love form. We love it in the universities with our various departments and disciplines. Likewise we love form and structure in industry and other institutions

Davis: Is Ford 2000 maybe what you are talking about?

Cole: No, it may or may not be. The whole idea is that you really focus on an objective or function and treat form as a variable in the true agile environment.

Davis: Sounds a little like the Manhattan project...

Cole: In a way I suppose it is, because there you had a consuming focus for what the objective was, and there was a sense of urgency about it that suppressed the importance of form...

Davis: And all the turfs.

Cole: That's right.

Martin: If you saw something didn't exist, you just went and did it.

Cole: And we have, historically. We did that kind of thing in the automotive industry routinely in the 50s — for example, the '55 Chevy with its new engine, transmission, body, chassis, the whole works. It was taken from concept to production in less than two years...and this was before computers.

Marks: With a very small team of people.

Cole: Of course. That was one of the key features of an excellent product development process.

Marks: And that is why this virtual company idea makes a lot of sense when you start talking about the diversity that the marketplace is looking for and the quick response time. Have you heard about bicycle manufacturing in Japan? You can go to your bicycle dealer and he measures you for fit. You decide what you want, all the custom features of your bicycle, and it will be delivered to your home in three days. They say, "We could do it in 48 hours but the customers wouldn't believe we really did it." (laughter)

The point is that people are looking for quick response time and more diversity and more ability to customize the product to their needs and wants. Increasingly, this is becoming possible if the upstream operation is configured appropriately. Again, the enablers are information and communication technology and some new organizational structures. When they are put in place they will provide a competitive advantage. That's what is going to drive it. The guy that can provide the quick response is going to get the market.

Davis: So we could conclude, as we look to the 21st century, the first order of change — because it is already on-stream or at the beginning of the stream — is going to be some very different ways of doing busi-

Dr. William R. Martin

ness, of organizing, not only business but campuses or what have you?

INSTITUTIONAL INERTIA

Martin: We (on the campus) are the last bastions to change.

Cole: You mean of resistance to change. One message we are suggesting here is that you can't look at things in separate packets because they are woven into a more complex mosaic and because of that, it makes it more difficult to conceptualize. Historically I think what we have done to simplify our world has defeated the idea of systems thinking. We have dealt with complex problems always but we have tried to break them up into parts and I think as a nation we have excelled at dealing with the parts. Look at the breakthroughs and the knowledge developed here at University of Michigan in the narrow disciplines. Throughout the entire U. S. I think we're really world class in terms of moving discipline-based knowledge forward. Where we have our problems is in taking packets of knowledge that have been separated from the whole and reassembling them back into the total system.

Davis: Any other macro issues?

Cole: I think there is the issue of government policy. The policy interface with technology cannot be ignored. Whether we're talking about government regulation, national defense or global competitiveness, the technology-policy interface is critical.

Another is the shift from a product-oriented thinking to a process thinking. Our traditional thinking was product focused where we thought about the end point. What we are moving to is a process mentality where we think of how we get to the end point. The problem we often faced with a product-oriented philosophy was that the process was not in control. If you think of Deming's philosophy and the idea of process control, normally we think of manufacturing processes. But now we are beginning to realize that it applies to everything else as well. There is a design process, marketing process, a process by which we create policy in government. We often look at the end point rather than how we get there. If the process is under control, the product of that process will take care of itself. If the process is not under control, you are lucky if you get the product you really want.

Another issue is the concept of what knowledge really is. What is knowledge? Can it be benchmarked and measured? We have attempted in many ways to use surrogates for knowledge. Organizations with large numbers of PhD's, MBA's, MS's, BS's say they are smart. Unfortunately a degree is not really knowledge itself, it is a surrogate measure of knowledge. Knowledge is really something else and there are several dimensions to it. One is knowing something and the other is knowing how to do something.— doing what you know, taking the knowledge or the intellectual capability you have and converting it into some kind of action. There are a great many smart people who know a lot but can't put it into a form of some value.

Martin: I would like to explore the information technology a little more too, a couple of items there. One is high-performance computing, advanced simulation along with a few things that you simply aren't able to do or take too long to do, for example simulating crash environments.

Virtual reality is another application of information technology that is going to revolutionize who knows what. We will see large inroads of this technology, including automotive design and many other areas. People who know these areas, who can apply the technology, are using them to design automobile interiors now. That's one obvious application.

Marks: It's in its infancy and it's on one of these (gesturing upwards) curves.

Martin: But clearly it works.

Marks: You're going to get to the point where the object can be viewed as a three-dimensional holograph, laser-created.

Davis: So virtual reality is an artificial simulation but in a controlled, total atmosphere, all in the computers?

Marks: Yeah, it's the ultimate three-dimensional visual representation. Right now we put the image on a flat screen or you must wear a hood. Eventually it will be a holographic image. For all intents and purposes it appears "here," only it is computer-generated. There is a huge advantage in the ability to modify that image and to interact with a three-dimensional virtual object, to have it simultaneously presented in more than one location. Many possibilities open up.

Martin: The advanced simulation techniques allow you to combine simulation with virtual reality, to be able to simulate the physical process while you tweak this or do that. Tweaking may actually have a huge effect somewhere that can't be predicted very easily. With high performance computing you can couple all that into the virtual reality.

Cole: Medical applications?

Martin: Yes, incredible medical applications.

Davis: One of the things I suggested is that, even though your disciplines are somewhat far afield, if we look back 100 years to the world of 1895, the average life span was 47 years and there were all kinds of health

Together, they revolutionized an industry.

In 1921, Henry Ford entrusted a young engineer named Jervis B. Webb with developing and installing the first rivetless chain conveyor.

From that point on, Jervis B. Webb Company has set the world standards for providing quality material handling systems design, engineering, implementation, and support.

Many of the world's largest corporations – as well as small to midsize companies – look to our full range of integrated systems and products to meet their material handling needs. They trust Webb to provide innovative, cost-effective manufacturing solutions that enhance both productivity and efficiency.

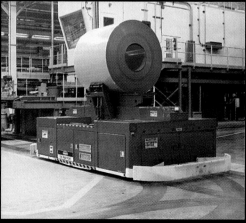

Jervis B. Webb Company WEBB
MATERIAL HANDLING SYSTEMS INTEGRATORS

34375 West Twelve Mile Rd.
Farmington Hills, MI 48331-5624 USA
1-810-553-1220   FAX: 1-810-553-1228

See us on the **Internet** http://www.industry.net/jbwebb

threats that were omnipresent. The key question today is the cure for cancer. We can't imagine what it might be, but if we're optimists we think, we're going to find it. We don't know if it will be next week or five years from now.

MULTIDISCIPLINARY MEDICINE

Cole: We know that human beings aren't immortal. We can eliminate cancer or heart disease but something else will then become the new medical challenge.

Marks: There is an interesting parallel here with some of the systems engineering and cross disciplinary kinds of things we spoke about earlier. The interaction of the medical community and the engineering community already is producing all kinds of neat new approaches. If you look at the incredibly sophisticated surgical techniques that are available today, they are not generated by medical people or engineers working alone.

Cole: Here at the University, the engineering college's ME's, EE's, even nuclear engineers are interfacing with the bio-medical or bio-engineering activities. These represent very significant activities within these departments.

Davis: So this is really a new marriage of engineering and medicine which hasn't really existed before?

Cole: No not really. When I was a student here the med school was actively recruiting the high-grade-point students to go into medicine — as physicians — and they wanted to bring engineering thinking into medicine.

Marks: The dentist who works in my mouth is building things on engineering construction principals and the materials that he's using have come out of developments in material science.

Martin: We hired a chemical engineering faculty member last year who has a joint appointment in the dental school.

Marks: The benefits of cross-disciplinary activity extend to a lot of fields.

Davis: If we are going to kill cancer let's get everybody in on the act, not just physicians. It isn't quite happening yet, but we can speculate it is becoming more common.

Cole: I think one thing we need to realize, and I will use information technology as an example, is that we're just starting up the learning curve. Most people today would say we are well up in the learning curve but in reality we are probably just getting started. If you recognize this, it is difficult to visualize the kind of ride in information technology we're likely to be taking over the next few years — but it will be an exciting one.

Davis: You can't predict what is going to be discovered.

Cole: We are moving from essentially an alpha numeric world to a visual world. Dan Atkins commented several years ago that a fairly small fraction of computing power was really being used at the interface between the human being and the computer. In the future most of that power will be applied to the interface, resulting in a much more friendly and powerful — computing environment.

Marks: Atkins, one of our faculty, believes we are headed for what is beginning to be called ubiquitous computing. He says, as technology continues to increase computer memory capability, speed, and sophisticated visual, aural and tactile presentation of information, we will eventually get to the point where we don't really know we have computers. They will become just a part of our everyday experience. Today, the interface between the computer and humans is still pretty clumsy, with keyboards, touch screens and its things like that. But as computer power continues to increase, he says that 95% of that increased capacity is going into improving the human-to-computer interface. The goal is to allow you to converse with the computer, to interact with it as though it weren't there. When you want something to happen, your natural way of communicating will make it happen.

Cole: We know now that they can tap into nerves, replace an arm, and people manipulate it on the basis of their thought processes.

Davis: Would that work to overcome paralysis?

Martin: Last night on TV, I saw a documentary on treating people with Parkinson's Disease. It was a little different but still they were able to localize a nerve area with Parkinson's Disease and block it. It was incredible.

Cole: We tend to look at that as medical but it really was technology. The whole idea of a visual world with a transparent interface to the digital world is profoundly important. You can say that things done by "cut-and-try" are basically in a terminal stage. The whole tradition of building something, or trying something, is moving to an analytic framework. For example, we have rapid prototyping technology like stereo lithography. It's a very clever thing where we shine a laser on a pool of light-sensitive polymers and essentially can "grow" a part. Is that virtual prototyping or is it just an intermediate step? Because probably we will not need to create that part physically but only virtually. From an engineering perspective what we're probably going to do is go from the concept to actual product. I think the potential for math-based or simulation-based engineering is to get to the point where we'll go

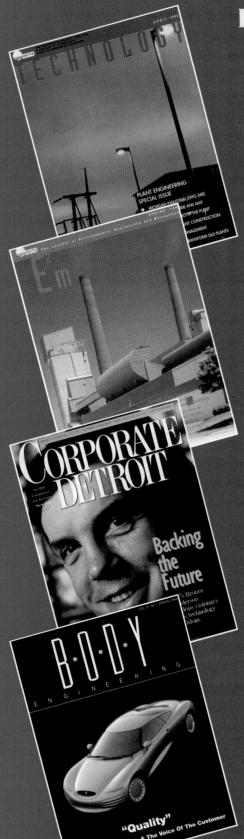

directly to pilot line production to validate what we plan to build. The whole process of engineering should be dramatically shortened. Essentially the role of the shop in that process almost disappears. The shop then becomes not the shop to create prototypes and to provide engineering things, but the shop that creates the production tooling.

Marks: Development of the manufacturing process is all going to be simulated.

Davis: You mentioned a couple of minutes ago that computing power is doubling ever three years, is it gaining momentum or losing momentum?

Martin: It is not gaining, it's still doubling, but over a longer period of time.

Davis: Because in the kinds of things we are talking about, I could visualize would just use huge amounts of capacity. But the huge amounts of capacity that we have today, were unthought of a few years ago.

Martin: I think we are seeing somewhat of a modest slow down of this rapid acceleration. So it is still accelerating but it's slowing down.

VIRTUAL EXPANSION

Cole: I think some people hope that is the case too. You have companies like SiliconGraphics. They did not exist 10 years ago and today they are a $2 billion company. They're in the forefront of the art of virtual reality and the visual representation of things. They must essentially reinvent themselves every year to year-and-a-half. This company and their peers are bringing some fantastic new tools to the world of engineering.

Marks: They are known for their work in Jurassic Park and Terminator II in the movie business, but their real impact will be in manufacturing businesses that will be changed by this three-dimensional visual computing in an incredibly important way.

Martin: There is a cut-throat feel to the whole field of computing with these start-up companies. The worst thing you can do with a start-up company is have a great product and not have another one, say 18 months behind, or you're dead. There is a whole list of companies that failed in the last 10 years, especially in the hardware business.

Cole: That's really what caught companies like IBM off guard. They sort of owned one paradigm and all of a sudden a new paradigm exploded on them and they were not sufficiently agile.

Martin: But their pockets were deep enough...

Cole: That's one of the nice things about an industry like the auto industry, nice in the sense you're dealing with large physical entities, complex things with many parts that weigh a great deal and have large struc-

tures. These kinds of things inherently have longer time scales associated with them than with computers, for example. So there is some comfort with a level of investment and hardware that slows the pace down a bit compared to the computer business.

Marks: In predicting the future, I think one of the stable things you can count on is the desire of people for personal mobility, the thing that has driven this automobile industry worldwide. Of course, with the tremendous changes of communication capability providing the ability for people to interact from a distance as if they were together, some impact on travel is bound to result.

Davis: Working from your home.

Marks: There's an interesting project today, tying together research people in four places in the world, running an experiment which is physically going on in Greenland. They all interact as though they were in the room with the equipment in Greenland, running the experiments.

Dr. Craig Marks

Martin: The University of Michigan offers a course with Cal-Berkeley that has the classrooms here and in Berkeley, and fully interactive students and faculty at both campuses.

Marks: So Dave's original theme of globalization says companies that are operating on a global basis need the ability to have teams interact in this fashion. That is beginning to happen now, not 10 years or 20 years from now.

Davis: Chrysler just built this huge, marvelous new technology center.

Marks: Which has a great advantage, to co-locate a bunch of people. Ford, on the other hand, is going to design cars on a worldwide basis. Ford 2000 envisions a system that will allow people, whether they are in Europe or Japan, in this country or in Southeast Asia, to interact as though they were in the same room. That technology will be a reality in the very near future.

STAYING MOBILE

So you ask then, relating back to my earlier statement, "How is this capability going to affect the demand for transportation?" I still think personal mobility is a very fundamental, human-nature kind of thing. Is radically improved communication going to change that? I believe it is going to supplement it. I don't think it is going to replace peoples' desire to go see Grandma for Thanksgiving.

Davis: Well, obviously production workers are still going to have to go to the plant. In theory, I suppose they could push a button even from home to operate the computerized work station. I was in the Ford Romeo Plant last spring and was astounded to see that when something goes wrong with a machine, it calls up its keeper on a beeper and says "come fix me."

Marks: The ideal is that it anticipates what is going to go wrong and calls before it shuts down.

Cole: That's prognostics. But the realities of what human beings are and what their needs are is critical. It is like trying to redefine a new architecture for a vehicle. One of things that is fixed is the architecture of the human being. Once you say a car is going to have to support five people and so many cubic feet of luggage, you define a great deal about the vehicle package. One of the things that comes with human beings is the need for social interaction. We have some very basic needs, so what technology does is help satisfy some of these needs, or give us more time for addressing what humans really want to do and less time dedicated to what they really don't want to do. So in a sense all that we're talking about here is a sort of a human enabler that can optimize the role of human beings in society.

When you think about information technology — and we talked about technology being way ahead of our ability to use it — one of the problems is that we have an information overlap. If you project that into an engineering environment, look at the vast amount of technical literature available. You don't want to spend a year to find a nugget pertinent to your job. So the ability to effectively manage vast quantities of information becomes crucial.

Marks: The information available electronically is exploding. Techniques are developing that will under-

stand what your needs are — and introduce what have been called agents, really surrogates for you, which will search through the maze of information — and provide back to you whatever you really want. These agents will search for whatever time and in whatever universe of information databases you are interested in. That is a remarkably effective way of getting information.

Cole: When you think of technology in general and our ability to use technology, things are not moving in a uniform state. We have some technologies such as information technology that have moved out in front of our ability to use them effectively. Life is a series of little steps and they are not all necessarily in sync.

Davis: We talked about the enabling processes that are ongoing, that we have in hand, or can reasonably project. The safest predictions are the ones that don't go too far. Can we shift over and try to think about some specific areas? Energy is one and I don't know what other broad things, because energy touches on so much and is so key to everything.

Martin: Currently, we're using hydrocarbons, a limited and finite resource, literally burning it up. We have a choice that we are not willing to implement.

Davis: Where do you think it is going to go?

Cole: Fundamentally, there is no energy shortage and there probably never will be, technically.

Davis: So it truly comes down to an optimum energy conversion. Hydrogen has infinite energy?

Cole: Well there is no such thing as infinite anything. But there is a lot of energy — for example, we could harness fusion. But when you look at hydrocarbon from a petroleum standpoint, it is quite inexpensive and is readily available.

Martin: Natural gas is cheaper. Gas plants are cheap, there is no financial risk.

CALIFORNIA QUESTION

Davis: What do we do about California?

Martin: California utilities are buying energy across the nation, so they are expecting to import their energy, export their emissions, and let the rest of the country worry about the consequences

Cole: Consider the zero emission vehicle program — they look at the use of the electric vehicle and don't regard the production of electrical energy as creating a pollution source.

Davis: There was much discussion of this yesterday by the retired engineering executives group at Ford Research Lab, along the lines of, you could take the biggest bite out of the Los Angeles pollution problem by getting the older cars off the road but local officials won't do it...

Cole: Or maintaining the cars better...80% of the problem is caused by 20% of the cars.

Davis: One of Ford's resident inventors, George Muller, years ago said that he thought we would end up with an electric vehicle. We have to have the way of getting there. We need a battery which hasn't been invented yet.

Cole: You could have an electric vehicle sooner than a hybrid, but a hybrid probably has greater potential over the long term. The hybrid can provide long driving range and negates some of the critical deficiencies of batteries, like relatively low energy and power density and the life-cycle problems with deep discharge. So it is, perhaps, a better solution, but it's a little further away.

Davis: It appears to be a cost issue and, I suppose, a pollution issue.

Cole: That gets to one of the fundamental drivers that will emerge, and that is, the reality or lack of reality of global warming as an environmental issue based on the burning of carbons. We're in a situation where we really don't know. There is evidence that suggests there is, and also that there is not. Some say, "so what if we do have a little global warming." Global warming could be a major consideration in how we view the environment. Pollutants like unburned hydrocarbons, nitrogen oxides and carbon-monoxide are really no big deal because it's a matter of economics as to how much we want to reduce them. Carbon dioxide is an entirely different matter.

An issue here is the fact we have reached the end of our regulatory rope, in being able to create wise environmental policies without having good analytical tools, things like economic analysis and risk analysis. We are certainly in an area with the car where affordability is a problem today, which means the economic issues related to the car are going to be much more significant than they have been. We must be careful not to add considerably to the cost of the vehicle for marginal environmental benefits. Very high levels of pollution control are probably going to be resisted by consumers.

ESD's panel discussion on the future. From left: editor Mike Davis, Dr. David E. Cole, Dr. William R. Martin and Dr. Craig Marks.

Davis: We in Detroit know this, but we haven't been very successful in convincing others. Now I understand that these government-industry teams are beginning to make a little headway.

POLITICAL PROGNOSTICATION

Cole: One of the keys to good forward thinking is to be able to "read the tea leaves." There was one big tea leaf laid out in last November's elections in this country which was a major political bombshell. Clearly one of the factors in the election is that people by and large don't feel they have been getting the greatest bang for their buck.

We have found in the car business, with a typical car being about $20,000, that the affordability issue is very important to consumers. They are concerned about value. One of the interesting parts of affordability is that consumers in many ways are their own worst enemies — because they are demanding things that they can't afford. In fact, if you look at a 1975 car, updated to 1995 with the same basic set of features as in 1975, it is more affordable than it was 20 years ago. In terms of the number of weeks an average American worked to buy that car, it is actually a little more affordable than it was 20 years ago. Now if you add regulatory mandates and the new features customers specify, it is a lot less affordable.

Marks: Mike, we haven't said anything very profound about energy except there is no looming energy shortage. The success of the automobile today owes a lot to developments in the petroleum industry and the work of many people who have honed the engine-fuel relationship over the years. As we go forward, this continuing evolution of engines and fuels could alter car designs and performance in a significant way.

Gibbs can provide
the following types of castings...

**Fan Clutch Body-
383 Alloy**
*Automotive Engine
High Balance Achieved
H. P. Vac. Die Cast*

**Cylinder Head Cover-
380 Alloy**
*Automotive Engine
Synchronously-Machined
at Gibbs
H. P. Vac. Die Cast*

**Valve Lifter Carrier-
380 Alloy**
*Automotive Engine
Three Years-Zero Returns
H. P. Vac. Die Cast*

Bearing Cap-HD-2 Alloy
*Automotive Engine
Near Zero Machining Scrap
H. P. Vac. Die Cast*

End Plate-356 Alloy
*Power Generator Motor
Permanent Mold Alloy
H. P. Vac. Die Cast*

**Input Housing-
390 Alloy**
*Automotive Transmission
High T.I.R. Requirement
H. P. Vac. Die Cast*

**Side Cover-Magnesium
AM-50 Alloy**
*Compressor
Superb Mechanical
Properties
H. P. Vac. Die Cast*
(Illustrated for discussion
purpose only.)

Front Head-383 Alloy
*Automotive Compressor
Zero Leaker Requirement
H. P. Vac. Die Cast*

**Connecting Rod-
380 Alloy**
*Small Engine
Zero Porosity Required
Ultra-High P. Vac. Die Cast*
(Illustrated for discussion
purpose only.)

Piston-390 Alloy
*Automotive Compressor
50% Stronger Casting
Solution Heat Treated
Vac. Die Cast*

Stator-380 Alloy
*Automotive Transmission
High Balance Achieved
H. P. Vac. Die Cast*

Clutch Piston-380 Alloy
*Automotive Transmission
Magnesium Modified Alloy
Ultra-High P. Vac. Die Cast*

Piston-380 Alloy
*Refrigeration Compressor
Super Porosity Control
H. P. Vac. Die Cast*

High Pressure Vacuum Die Casting —
A casting formed when metal is injected into a cavity containing no air under pressures ranging from 2,000 psi to 10,000 psi. Such castings may be subjected to temperatures of up to 850° F without blisters.

Ultra-high Pressure Vacuum Die Casting —
A casting formed when metal is injected into a cavity containing no air under pressures ranging from over 10,000 psi to 20,000 psi.

Solution Heat Treated Vacuum Die Casting —
A die casting that has been solution heat treated to enhance its performance properties. (Castings equivalent to Squeeze Castings, in general.)

**Permanent Mold Alloy
High Pressure Vacuum Die Casting —**
A casting formed when a permanent mold alloy is injected into a cavity containing no air under high pressure. This permits new opportunities.

**Magnesium Alloy
High Pressure Vacuum Die Casting —**
A casting formed when a magnesium alloy is injected into a cavity containing no air under high pressure. Magnesium alloys cast with our process have super elongation.

Machined & Assembled Casting Products —
Synchronously finished die cast components and assemblies

Gibbs
Die Casting Corp.

369 Community Drive • Henderson, KY 42420 • (502) 827-1801 • FAX (502) 827-7840

Davis: You're speaking of fossil fuels or hydrocarbons?

Marks: Traditionally, we have worked with variations of hydrocarbon blends, but at the SAE meeting last month, dimethyl ether was revealed as being a good fuel for diesel engines. It was reported that the economics may not be too bad, and this fuel produces a completely different tradeoff between NOx and particulates. There were four papers, which involved a Danish firm in the petroleum business, Haldor Topsoe, and an Austrian firm, AVL, who had done a lot of the basic research work.

The initial supply could be obtained by converting a plant which is building methanol today. The interesting thing about this discovery is that, forever and a day, it was chiseled in stone that as you decreased NOx in a diesel, emission of particulates increased and visa versa. This fuel, dimethyl ether, simply eliminates that unfavorable trade-off.

Cole: It is a paradigm shift fuel for diesel.

Davis: Some rebirth of the diesel possible?

Marks: The diesel is very much alive and kicking in Europe. Some people promote the diesel as a "green engine" in Europe today. The reason is, its fuel consumption is lower. Carbon dioxide is regarded as more of a problem in Europe, and people worry about the greenhouse effect. The broader issue on this subject of fuels is our ability to create new molecular structures with predetermined physical characteristics. This applies in the materials field as well. We are beginning to be able to specify what it is we want the final material to do, and build the chemical compound and structure to produce that result. This is happening in drugs, in some petroleum products and in some basic materials, particularly polymeric materials. The ability to have entirely new levels of performance for materials or for fuels is something that is on the horizon. The enabling technology again goes back to simulation and computing power, the kinds of things that we talked about earlier.

Martin: Without having seen virtual reality, there is no way you can predict anything to do with it...we're simply not on the same paradigm.

Davis: With our wonderful 20/20 hindsight, if you went back to 1895, and you had the kind of knowledge then that you have today relatively speaking, where do you think you might have come out? Look at Scientific American at the turn of the century, the lead article's all on vehicles, horseless carriages, the whole thing's on steam.

Cole: Look at Jules Verne, look at what a visionary he was. I don't think he was necessarily a visionary because of substantive knowledge. I think he was guessing and dreaming. I think unfortunately it is hard for engineers steeped in technology to really dream.

Marks: I am beginning to think that the pace at which change will take place has less to due with technology today than it has in the past. Technology is moving so fast that it is outpacing our ability to assimilate it. The rate of change has much more to do with infrastructures and organizations and the ability of people to adapt to change than with the availability of technology. Another issue arising from this rapid pace of technical change is how to increase the technical literacy of the general population. For progress, government policy needs to be supportive, and in a democracy you are paced pretty much by an infrastructure that can be supported by people's understanding of what is going on. If people are 20 years behind the technology, you are not going to get that technology to come to full bloom very rapidly.

TEACH ABOUT TECHNOLOGY

What this suggests is that we must address how to teach people about technology, how to get them to understand what benefits it offers. Again the digital world is making a lot of educational innovations possible, self-paced education and hands-on exploratory kind of education on a life-long basis. We are going to get to a point where people can, just from their own curiosity, learn things you can't begin to teach in a formal setting. It has been said, with a computer and a CD disk you can teach physiology today to kids in high school better than we could ever teach it in college with a professor lecturing. I'm not saying there isn't a need for formal education. What I am saying is, there's a tremendous need to get people to understand more about technology and its proper use, in order for us to reap the benefits available today.

Davis: In the real world we're looking at what it seems to me a dumbing down. Am I not correct that college board scores in science and math are lower than they used to be?

Cole: The real paradox we're dealing with is that we're far more sophisticated with technology, it is far more comprehensive and pervasive.

Now a corollary to that, again related to systems thinking, is the blending of the behavioral and social sciences with the hard sciences. You can see some clue to that if you look at who runs organizations. People who run organizations such as the car companies are typically either from the business disciplines like finance or accounting, or from engineering. Characteristic of these people is that they are very quantitative.

Show them a number and they can make a decision. Many went into business and engineering because they didn't like the social and behavioral areas. But when you look at what the impediments are to change, or the transformation of industry, or more relational, cultural types of things, that's all over on the "soft" side. If you think of a box as an organization, in part of it is the "hard" or quantitative stuff, this is where most of us live. In the other part of the box are the "soft" issues, the people issues, the qualitative. Quantitative people are not very skilled in this part of the box.. We need to manage the interface between the hard and the soft. Where are the soft people, the behavioral scientists or the social scientists? Rarely do you find them inside technically-based organizations. One of the things I can envision is the movement of behavioral, social-science-based thinking into our technically oriented organizations, leading to a better blending of the hard and soft sciences.

Davis: In the behavioral sciences, I can see a very good case for virtual reality. That was very clear from what I saw yesterday, where they were physically imposing additional tasks on the driver so they could measure performance.

Cole: How do you motivate people to change, how do you motivate people to learn, how do you motivate people to accept a different environment. From the hard side, what do we say, "Either you change or else"? Soft thinking really focuses on the total human being.

Martin: An example of how we have failed miserably in educating the public about science and technology in general is nuclear power, particularly nuclear waste. When you talk about encapsulating nuclear fuel, waste disposal, it scares people to death. The problem is politics and law, not technology. You have to take care of a modest amount of heat, ten 100-watt bulbs is the amount of heat to dissipate. You want something that won't corrode for a hundred million years? Use an inch of copper, capped with stainless steel on the outside. Yet we have people perceiving this overwhelming "national problem" that is absolutely not there.

Davis: It sure is an education issue. But I think it has to do with the American Psyche. We are a nation of radicals because our ancestors came here because they were going to get out of the old country. It is a fear of technology.

RELATIVE RISK

Marks: It's something even more fundamental than that. It has to do with the idea of zero risk. Some public policies say zero risk is the only acceptable amount.

This is another public education issue, because there is no such thing as zero risk for most activities. But, until you understand the concepts of probability and statistics, you are not going to believe that.

Davis: They only apply that to hard science. They don't apply this zero risk thing to their lives.

Cole: There was an interesting TV program a year ago. The moderator had a group of people in a room talking about environmental risk. One vignette addressed a city in Colorado where many people had been living in homes built on mine tailings for over a 100 years. The EPA declared it a super fund site and wanted to move the town and haul the stuff away. Somebody in the town said, we've been here 100 years, let's see how sick we are. So they conducted epidemiological studies and found there were no health problems. Now EPA is so insistent there is a problem, they are force-feeding pigs dirt from the town to see if they can find some health problem. On the same program the host said to the audience, what if we had this new, very clean fuel, but there are going to be 15 people killed next year because of its use. He asked the audience whether they would accept this fuel, and they rejected it. Then he said, well, we have that fuel today, it is natural gas, and there are something like 200 deaths a year related to natural gas. Everybody was horrified.

This is a perfect illustration of the zero-risk mentality that we have. It is really a curse in terms of effective decision-making, whether we talk about asbestos problems in the schools or air pollution. My point is that we are moving in the right direction when we talk about affordability and risk and economic analysis. We must use science to facilitate practical decisions.

Davis: Truly you are preaching to the saved, but my observation is we have a huge barrier in the school system because I suspect that our biggest problem is teachers at the elementary and high school level. Not to say there aren't some very good teachers out there.

Marks: There are some exciting experiments going on and some school systems are doing very innovative things. But many schools are still doing things the same old way and I am told that parents, in general, are very satisfied with their schools. This may be good or may be bad. People need to understand more about how much better the educational process could be, by using some of the innovative approaches in these experimental programs.

Davis: Certainly computers have invaded the schools.

Marks: It is more than just computers and technology, however. It is understanding very different ways for people to learn. Bill, why don't you describe what is going to happen in this new instructional center being built on the North Campus?

Martin: The Integrated Technology Instructional Center is a forty- million-dollar state-funded facility. It's actually hard to define its use, it was intentionally built with the idea that we don't really know how it's going to be used. It's going to evolve as our needs change. It's going to be a place where engineers, artists, musicians meet in the middle of our campus. It's across from the architecture building, across from the music building, and right next to the engineering building. It will have performing studios and then it is going to have our digital library, wired to the world. There is a place there for a virtual reality lab.

EXPERIENTIAL EDUCATION

Marks: Students will be teaching students as much as professors will teach students, and people are going to learn in a experiential way. I think that kind of learning is going to revolutionize the educational process.

Davis: What you are saying is by taking people that are in the soft disciplines and exposing them to technology it may open a door to technology that wouldn't have been there before?

Marks: We tend to separate left brain and right brain education. For instance, we teach engineers to be analytical people. We sort out kids from the time they are in kindergarten-first grade, by whether or not they are good in analytical areas, and encourage those that are good to go into technical areas. Creativity is a totally different matter than analysis and the creative artist uses a totally different approach to solving problems. But a good engineer often needs to be very creative. Melding these skills by broadening the educational process is what is going to happen in this center.

Someone described K-12 education as this wonderful process where we take children 5 or 6 years old who have an unbounded interest in learning and experiencing everything, who perceive everything they touch as something new to be learned, and in 12 years we turn them into robots that resist learning, so that you have to cram down their throat every piece of learning they get. By the time they graduate, you have to force feed them. That's dumb. There must be better ways and, yes, I believe the computer is an enabling technology which can help some of that happen.

There are other process changes being tried, too, such as ungraded classrooms, team teaching and team learning, and around-the-year schools without long summer recesses.

Davis: If it works and spreads, are we going to get rid of our agriculture-based school year?

Cole: Industry now realizes it has to optimize its use of resources. If you have a factory not working on weekends or evenings, that's a poor use of those assets. We must use all of our assets and resources effectively and efficiently, including plants, people and natural resources.

Davis: Globalization will get to us, I guess, on this.

Marks: The fact is, today we are not competitive in our educational process.

Davis: Maybe we have identified a major area of change which has no technological aspect to it.

Cole: There are many fuzzy boundaries here. Some things are very difficult to compartmentalize. Our traditional way of organizing things is to break it down into its parts, that is how we write papers, conduct research, do engineering...

Martin: It's how we teach school....

Cole: ...and maybe we are going to have to rethink that process. Now we are looking at some tools that can enable that.

Let me throw out another tremendously important issue, the idea of values, ethics and morals. Where does that fit into the context of what we are talking about here? We talked about human beings, collaboration, and new ways of thinking, but we have some real problems. Many think there's a "value" world and then there's the separate world of technology. The fact is, these worlds are becoming intertwined.

If you consider ethical issues in industry, you have two things that have happened in the past few years. One is the fast pace of change with a high rate of turnover of people. The embedded ethical code that was part of the history of industry, and was passed from one to another in an orderly fashion, has begun to be fractured by the speed of change.

Secondly, with globalization we have now the intermixing of many cultures at a very high rate. Many of these cultures have different ethical codes. Both factors are creating problems. We are restructuring to work in more interdependent ways. Companies and individuals are trying to work together more effectively. We believe in teams and close working relationships. One of the things fundamental to achieving close working relationships between groups, individuals or teams is trust. We have to trust one another. What is the foundation of trust? It's a shared value system. You cannot have trust if you're working with one code and somebody else is working with another code.

Davis: That's a very interesting thought. I could see how that works, because it is so value-related, and it's the breakdown of some of those traditional values that's got us in some of the pickles we're in from a societal point of view.

CULTURAL CURVEBALLS

Cole: A perfect example is that in a number of cultures, kickbacks are standard business practice...

Davis: We're practically the only innocents here in the United States... The whole substance of our conversation this afternoon strikes me as very humanistic. You're technologists and yet this is very humanistically based and somehow I don't think that would have been a consideration 100 years ago.

Marks: That may be why some technologists have made predictions that don't look very sharp 20 years later, because they didn't take into account the human nature side of things.

Martin: Have we wrung it dry or hardly started?

Cole: I don't think this is a very easy task to tackle.

Davis: I didn't expect it to be. I didn't really want to come forth with, well, we're going to have an all-plastic car that will get 100 mpg by 2003.

Cole: Good. I think what we are suggesting is, there is a whole different thought area that is non-traditional for technical people to deal with. Certainly it's reflected here in our organizations that want to be very quantitative, "show me a number and I'll make a decision," and realizing there is a whole universe of things you can't deal with in that fashion. You can talk about values, ethics, feelings, relationships and things of this type. The agile environment, virtual organizations, all of these things require a kind of a blending of what we are talking about. The behavioral, life and social sciences, the hard or traditional physical sciences, they are all becoming connected. This will be a great challenge for all of us.

Contributors to

The Technology Century

Bruce J. Annett Jr. *is director of university relations and alumni services at Lawrence Technological University which he joined in 1976 as director of public relations. Earlier, Annett held positions with DePauw University and Albion College. He is a director of the Oakland County Pioneer and Historical Society, and a member of the International Association of Business Communicators, Engineering Society of Detroit, Economic Club of Detroit, among others. Under Annett's leadership, Lawrence Tech has won a large number of awards for publication excellence. He holds a B.A. from Albion College and an M.A. from Michigan State University. A Michigan native, he was born in Pontiac and grew up in Waterford Township, Mich.*

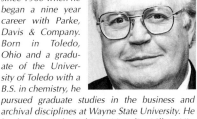

Janet Braunstein *joined The Associated Press in Miami out of the University of Florida and was transferred two years later to Detroit as AP's national auto writer in 1986. In 1988, she moved to the Detroit Free Press, where she was an automotive writer and then a marketing columnist before striking out as an independent journalist in 1994. Braunstein currently specializes in automotive writing for technical trade and consumer magazines. She lives In Grosse Pointe, Mich.*

Frank Buttler *is a Troy-based public relations consultant, a pursuit he began in 1970 after having been director of development for the Detroit Institute of Technology and communications director for the Greater Detroit Chamber of Commerce. For 12 years he was with American Motors Corp., where he was responsible for the appliance division's public relations. A journalism graduate of Ohio State University, he has recently earned credit in a master's program at Central Michigan. Buttler has been a longtime member of the Society of Professional Journalists, the Public Relations Society of America, and was a member of the ESD Publications Committee for many years including as chairman.*

With dual loves of engineering and history, **Bob Casey** *has a mechanical engineering degree from Rose-Hulman Institute of Technology, a bachelor's degree in American history from Towson State University, and a master's degree in history of technology from the University of Delaware. He worked for 12 years as an engineer at Bethlehem Steel's Sparrows Point, Maryland plant, and has been a historian and curator for the Institute of Electrical and Electronics Engineers, Sloss Furnaces National Historic Landmark, and the Detroit Historical Museum. Bob is currently curator of transportation at Henry Ford Museum & Greenfield Village.*

Bob Cosgrove *has been a Detroiter since 1960 when he began a nine year career with Parke, Davis & Company. Born in Toledo, Ohio and a graduate of the University of Toledo with a B.S. in chemistry, he pursued graduate studies in the business and archival disciplines at Wayne State University. He has been a resident of Detroit's Indian Village Historical District since 1965. He is a founding member and a past president of the Bluewater Michigan Chapter, National Railway Historical Society; a member and since 1989 associate editor and trustee of the New York Central System Historical Society; and in 1973 he founded the Indian Village Historical Collections, of which he is a past president.*

Charlotte W. Craig *is a business writer for the Detroit Free Press, where she has worked since 1979. After stints in the city room and the City-County Bureau, she was named an assistant city editor in 1983. Craig joined the Free Press Editorial Board in 1990, and served as reader representative and assistant to the executive editor until going back to the business staff in 1994. Before the Free Press, Craig worked for three years in Ford Motor Co. public relations. Before Ford, she was a feature writer and columnist at The Detroit News. A graduate of the University of Missouri School of Journalism, she lives in Grosse Pointe Park, Mich. She is a member of the Economic Club of Detroit and is a past president of the Detroit Press Club.*

Mike Davis *(editor-in-chief)* is author of architectural and marine history books and countless newspaper and magazine stories. An adjunct professor of architectural history at Eastern Michigan University, he formerly was Detroit Historical Society executive director, Evening News Association corporate communications director, Ford Motor Company technical information manager, Business Week assistant editor and a Miami Daily News reporter. With a B.A. from Yale and doctoral studies in technological history at Wayne State University, Davis belongs to the Automotive Press Asssociation,, Society of Automotive Engineers, Detroit Section SAE History Committee, Public Relations Society of America and the Algonquin Club.

Chris Dyrda currently chairs ESD's Publications Committee, playing a key role in shepherding The Technology Century. With ESD since 1988, he has also participated on its Membership Committee. Dyrda is manager of vehicle programs and control in Chrysler's Minivan Platform, as an outgrowth of his 25 years of engineering experience in vehicle designa and development. A Troy resident, he holds a bachelor's in electrical engineering from Michigan State University, and a management degree from the University of Michigan.

For the past 15 years **Paul A. Eisenstein** has operated The Detroit Bureau, an international, multimedia news service specializing in automotive, high-tech, bus-iness and related topics. His work appears in such publications as The Christian Science Monitor, The Economist, Investor's Business Daily, Parade magazine, Germany's Auto Motor und Sport, and Nikkan Jidosha Shimbun, Japan's leading automotive publication. He also appears on a variety of radio and television networks. His most recent among a number of awards is the Wheels award, presented by The Detroit Press Club Foundation for coverage of the nascent electric vehicle industry.

Al Fleming joined Eisbrenner Public Relations, where he is now vice president, following more than 35 years as an automotive reporter, columnist, editor and publisher. Fleming spent more than 12 years as an editor at Automotive News. Before that, he was executive editor of Ward's Auto World, and has been automotive columnist for the Christian Science Monitor, the Detroit Free Press and other publications and media. He is author of the book "Adventures in Autoland." The Bloomfield Hills, Mich. resident also has been heard daily since 1972 on WWJ radio's "Al Fleming's Automotive Insight." His awards include the American Business Press Jesse H. Neal Award for best series of editorials and feature writing. A native of Detroit, he is a graduate of Wayne State University.

James P. Gallagher is a 40 year observer of the construction industry and architecture, as a writer and editor, author and public relations executive. A senior editor for 18 years for Time, Inc and McGraw-Hill on their housing magazine, House & Home, he returned to Detroit in 1970 and served as director of public affairs for the architectural/engineering firm of Smith, Hinchman & Grylls Associates. He is the author of the history of that firm's first 125 years, and of a history of the origin and the restoration of the Wayne County Courthouse. Now retired, he has renovated nine houses in Detroit's historic West Village. He continues to write for various magazines and newspapers.

A local journalist since graduating from Wayne State University in 1977, **Martha Hindes** spent 10 years reporting news, business and feature stories for The Detroit News. In 1987 she formed Automotive Bureau to cover industry-related topics for local and national publications. A member of Detroit's Automotive Press Association, she also served for several years as a celebrity judge at the Detroit Windsor International Freedom Festival's Wheels of Freedom collector car contest. She has been a guest lecturer on feature writing for WSU journalism classes.

Charles K. Hyde is professor of history at Wayne State University. He earned his bachelor's in history from the University of Massachusetts at Amherst and his Ph.D. in history from the University of Wisconsin at Madison. He has written several books, including Detroit, an Industrial History Guide, (1980); Old Reliable, An Illustrated History of the Quincy Mining Company (1982); The Northern Lights: Lighthouses of the Upper Great Lakes (1986); and Historic Highway Bridges of Michigan (1993). He will soon publish a history of the U.S. copper industry.

Daniel Jarvis is a communications specialist with Detroit Edison. A former newspaper writer, photographer and production manager, he is a 1987 graduate of Wayne State University and serves as vice president of the WSU Alumni Association Board of Directors.

Mike Kollins has had a long career as an automotive engineer, historian and writer — and race car driver. An Ohio native, Kollins worked as an automotive engineer for Chrysler in the 1930s and '40s before serving in the Navy during World War II. He then worked for Studebaker-Packard in the '40s and '50s before rejoining Chrysler in 1955. He has contributed to Motor Trend and to auto maintenance publications and has held memberships and leadership positions on a number of professional boards and associations, including Society of Automotive Engineers, U.S. Auto Club and the Michigan Industrial Training Council.

Reginald Larrie, a Michigan native, received his bachelor of arts from Upper Iowa University, his master's in education from Marygrove College and a Ph.D. from Pacific Western University. His  byline has appeared in numerous publications throughout the country, including The Michigan Chronicle, Detroit Free Press, the Detroit News and Ward's AutoWorld. Larrie has also taught courses in Germany as a visiting professor of African American History for Wayne State University and is an instructor at Wayne County Community College.

Jon Lowell is a 25 year veteran of Detroit journalism. He is currently auto editor at WXYZ-TV, a commentator on WJR radio and a free-lance writer. He spent 12 years as a Newsweek correspondent based in  Detroit, 10 years as senior editor at Ward's Auto World, and three years during the early 1980s as director of communications for Burroughs Corp.

Richard E. Marburger is president emeritus and professor of physics, mathematics, and computer science at Lawrence Technological University, where he joined the adjunct faculty in 1965. He served  as president of the university from 1977 to 1993, when he returned to full time teaching. Marburger is past president of the Engineering Society of Detroit and other distinguished groups. His 40-year scientific career includes time at GM Research Laboratories, at Lawrence Tech, and in the Air Force, where he collaborated on the development of a photovoltaic cell. A native of Detroit, Marburger earned three degrees in physics, including his Ph.D., at Wayne State University. He won ESD's Horace H. Rackham Humanitarian Award and the ESD Affiliate Council's Gold Award, among many other honors.

Hugh W. McCann, a native of Northern Ireland and a graduate of Indiana Institute of Technology and the University of Michigan, covers science and technology for the The Detroit News. A former engineer, he  began his professional career in journalism at Newsweek magazine in 1959 and reported for the Detroit Bureau for seven years. He worked as a general assignment reporter for the Detroit Free Press from 1969 to 1976. McCann has produced television documentaries, authored a novel set in World War I and co-authored a book on World War II.

Local free-lance journalist **Maureen McDonald** writes business articles for the Detroit Free Press and various automotive publications, teaches corporate newsletters at Wayne State University and pro-  duces newsletters for several clients. Over the last 20 years she visited car dealerships across the U.S. while writing for a Chevrolet training publication, wrote automotive stories for the New York Times and edited several automotive fact books for Ward's Communications. She also did stints in PR at the Southeast Michigan Council of Governments (SEMCOG) and New Detroit.

Louis Mleczko has been a staff writer for The Detroit News since 1971; as transportation writer there, he broke the story on the Zilwaukee Bridge accident of 1982, and reported on construction of  the Detroit People Mover and its design and construction problems, among other metro issues and beats. Prior to that he spent four years at Akron (Ohio) Beacon Journal, where he shared the Pulitzer for coverage of the shooting deaths of four Kent State University students by Ohio National Guardsmen in May 1970. He also worked at The Macomb Daily, where assignments included 1967 Detroit riots and 1968 World Series. He is a graduate of Monteith College, Wayne State University, 1969, Bachelor of Philosophy degree, journalism major.

Having moved to magazine publishing from daily and weekly newspaper editing and management, **David Tell** is managing editor of ESD's Technology and E2m magazines and designed and copyedited its cen-  tennial book. Leapfrogging from brief stints with his alma mater Brown University's arts and literary magazine and with the Lowell (New Hampshire) Sun, Tell became copy editor and food columnist/restaurant critic for the Worcester (Mass.) Telegram. Upon moving to Michigan in 1990, he joined The (Adrian) Daily Telegram as assistant news editor/business editor, later becoming news editor for SCN Communications Group's weekly newspaper and its monthly Lakefront magazine in Oakland County. Tell's background also includes graduate work in philosophy and several years' employment in human resources.

Donald Thurber has pursued a long career in public relations and fundraising management since graduating Harvard in 1940. He now serves as honorary chairman and senior consultant to  the firm he started in 1958 and from which he retired in 1982. He has been a member of the Michigan Crippled Children Commission, the state Board of Education and the Michigan Historical Commission and was a regent of the University of Michigan and first chairman of the Board of Trustees of Wayne County Community College. His principal activities since 1986 have been on the board of Blue Cross Blue Shield of Michigan, as chair until 1992, and presently, as vice chairman of the Michigan Historical Center Foundation.

Gary S. Vasilash has worked in advertising and journalism since getting his master's in literature from Eastern Michigan University in 1977. He has held positions as senior editor of Manufact-  uring Engineering, associate editor with The Rockford Institute and creative director with G. Temple Associates. Now editor in chief of Production magazine, Vasilash has also been published in The Wall Street Journal, Lightworks, Werkstatt und Betrieb and elsewhere.

J. R. Wargo *is a consultant to industry on public interest activist tactics and agendas. He received a bachelor of arts degree from a Midwestern university in 1956. A six-year U.S. Navy career included*  *tours as a radar officer with blimp and fixed wing patrol squadrons, the study of Mandarin Chinese at the Naval Intelligence School, and service with the 7th Fleet in the North Pacific. He was UPI bureau chief in Cleveland, then Detroit bureau chief, McGraw Hill World News, 1965-1973. Following a professional journalism fellowship at Stanford University, he worked in Washington, D.C., covering federal regulatory agencies for Business Week. From 1976 until 1991, he was associated with a trade association representing the nuclear industry. He resides in McLean, Va*

Doug Williams *has as a consequence of three decades as a journalist gained much more appreciation for history as a discipline and endeavor than when he was a student at Wayne State University. A onetime editor of ESD publica-*  *tions, Williams has also done time at Iron Age and the Detroit Free Press. A resident of Grosse Pointe Woods, Mich., he has written and published about technology, business, cars and crime.*

Natalynne Stringer Williams *is marketing director for the Detroit Zoological Institute and a freelance writer/editor. Formerly, she was marketing and PR director for the Detroit Historical Department, an*  *account executive at Casey Communications Management, communications manager for the Southeast Michigan Council of Governments, and a PR representative at Coca-Cola USA. Williams has also done a number of free-lance editing projects for Gale Research, Inc. and has worked as a publicity/marketing consultant; clients included the Detroit Institute of Arts, the Metropolitan Detroit Convention and Visitors Bureau, the University Cultural Center Association, and the Museum of African American History. Williams is a graduate of Wayne State University.*

Anthony (Tony) Yanik *is editor of the National Automotive History Collection publication, "Wheels," and has authored numerous articles on automotive history for this publication as well as others such as the*  *"Chronicle," the bimonthly magazine of the Historical Society of Michigan. He is a member of the Society of Automotive Historians, Inc., the Algonquin Club, and the Society of Automotive Engineers Historical Committee, for which he organized and chaired a special session honoring five automobile designer greats in 1990, and presented the paper, SAE92085, "The Automobile: Unwanted Technology — The Later Years" in 1992.*

Tom Yates *is a freelance journalist living in Inkster, Mich. In the past he's been a telephone operator, power house operator, layout artist, temporary worker for Kelly Services, racing program coordinator*  *for an auto racing series in the mid-'70s and pursued another dozen or so other occupations. He's been free-lancing for 15 years, the last six full-time. His work has appeared in both the Detroit Free Press and Detroit News, "Popular Mechanics," "Autoweek," "Motorhome" and more than 15 other industry, automotive and general interest publications. He's one of the few members of the Detroit Press Club and APA who skydives.*

About Who's Who

The following sponsorship-advertisements, as well as the preceding advertisements in this book, illustrate the depth and breadth of support that ESD — The Engineering Society has enjoyed from the leaders of industry and the pioneers of technology.

We ask that you pay particular attention to the companies who are represented here. Among them are some of the oldest companies in America. Others are newcomers, representing technologies not dreamt of when ESD was begun. Many from both groups reflect the nature of today's global economy.

While we hope that all of these companies survive the next one hundred years, we know that will not happen without the vigilance and support of organizations like ESD — The Engineering Society, whose mission is to advance the field of education for the benefit of all.

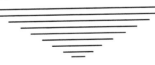

 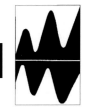

251

261

On behalf of the staff of Engineering Technology Publishing, Inc., the official publisher for ESD, the Engineering Society, we extend our thanks to all of the companies which joined ESD in its centennial celebration.